BEHIND THE WHEEL AT CHRYSLER

DORON P. LEVIN

BEHIND THE WHEEL AT CHRYSLER

THE IACOCCA LEGACY

HARCOURT BRACE & COMPANY

New York   San Diego   London

Requests for permission to make copies of any part of the work
should be mailed to: Permissions Department, Harcourt Brace & Company,
6277 Sea Harbor Drive, Orlando, Florida 32887-6777.

Library of Congress Cataloging-in-Publication Data
Levin, Doron P.
Behind the wheel at Chrysler:
the Iacocca legacy/Doron P. Levin.
p. cm.
Includes index.
ISBN 0-15-111703-9
1. Chrysler Corporation—Management.
2. Automobile industry and trade—United States—Management.
3. Iacocca, Lee A.   I. Title.
HD9710.U54C469   1995
338.7′6292′0973—dc20   95-7914

The text was set in Bodoni Book
Designed by Linda Lockowitz
Printed in the United States of America
First edition
A B C D E

*This book is dedicated with love to
Adina, my wife, partner, and counsel
of twenty years.*

CONTENTS

FOREWORD

NO MACHINE has shaped American life or enhanced American economic strength as the car has. The car bestowed mobility, privacy, and freedom on a generation that never dreamed such things were possible, and, later, on succeeding generations that took them for granted. Automotive factories became workplaces for millions and the source of middle-class incomes. These same factories supplied the armaments that won World War II. Highways, shopping centers, drive-in movie theaters, the rise of suburbia, the decline of cities, migration to the Sunbelt—all of these phenomena owed their existence to the actions of industrialists from Detroit.

Not surprisingly, the workings of the American auto industry and the activities and thinking of its leaders have been subjects of intense scrutiny. Detroit, after all, was the birthplace of both the assembly line and the nation's mightiest trade union. The city was the adopted home of Alfred Sloan, who practically invented business management. U.S. automakers gave the nation a managerial class and an endless number of technological innovations. Toward the end of this century, a challenge to the U.S. auto industry's supremacy by Japan, a nation humbled by the United States in World War II, called into question fundamental assumptions about the soundness of American manufacturing, about the quality of American business management, and about the principles of free trade and the relationship of government to industry.

When the Chrysler Corporation hovered near bankruptcy in 1979 and 1980, the nation's dependence on the vitality of large manufacturing entities became chillingly obvious. Chrysler wasn't just another company but a key component of the nation's economy. Rugged free-market capitalists favored letting Chrysler undergo a court-supervised reorganization. It might have worked, though the idea proved politically unfeasible: too many municipal governments, jobs, banks, union locals, schools, churches, and car dealerships depended on healthy sales of Chrysler cars. The federal Loan Guarantee Act that helped Chrysler survive reflected a political consensus and a national will to avert the failure of so dominant an economic institution as a Big Three automaker.

The events surrounding Chrysler's brush with failure elevated the automaker's chairman, Lee A. Iacocca, to a position of unprecedented national prominence for a business leader. Prior to 1979, Iacocca was known mostly for his celebrated firing by Henry Ford II from the Ford Motor Company. The loan guarantee debate, Chrysler's subsequent return to health, and the publication of his best-selling autobiography conferred mythic status on him as the nation's economic Winston Churchill. At the peak of his popularity, many Americans believed not only that Iacocca held the answers to the nation's economic ills but also that he should lead the country as president. Iacocca's platform consisted of one main plank, that the government had mishandled trade and the economy, precipitating a crisis for Chrysler and other companies. Ultimately, Iacocca heeded those who warned him that business executives made miserable politicians. But he seized a central position in the national economic debate, preaching his ideas in a weekly newspaper column, in speeches to gatherings such as the American Bar Association, and in Chrysler advertising, which, through clever use of patriotic imagery, reinforced his advocacy of "getting tough" with Japan.

Like many great leaders, Churchill included, Iacocca proved much better in war than in peace. When Chrysler was floundering, he marshaled resources, sustained morale, and decided swiftly on effective tactics to raise capital and sell vehicles. After Chrysler returned to health, Iacocca—like many of his peers in the automobile industry—failed to recognize and react to the vast, fundamental changes in the nature of global economic competition. The world of commerce had been one kind of arena following victories in two world wars, when the rest of the world lay prostrate; within a decade, the world turned into quite another, more difficult sort of place to do business, as cheaper (and sometimes better) products began arriving from abroad, heralding new and innovative practices in factories, in engineering laboratories, and in executive suites.

At first, most American business leaders were, quite understandably, in a state of denial. It wasn't *their* fault that big companies seemed unable to compete at home, never mind abroad. Few were ready to believe that foreign companies held advantages other than cheap material, cheap labor, and generous subsidies from their governments. The notion that Detroit or Pittsburgh or San Jose had anything to learn about management seemed laughable. Yet leaders of the American steel industry, by failing to invest adequately in new technology or to forge progressive labor practices in their factories, unquestionably hastened the debilitation of once-great manufacturers like U.S. Steel. Too little thought to technological research wounded U.S. electronics manufacturers as well. Thus, Sony and Toshiba and Panasonic inexorably replaced Westinghouse, Admiral, and other once-ubiquitous names as brands of choice for buyers of stereos and television sets.

The rise of the Japanese automobile industry was hardly a sneak attack. From the 1960s on, Detroit had plenty of time and a big picture window to study the vehicles Japan was producing and the operation of their factories. A plane trip to Los Angeles was all it took to see firsthand the growing number of Japanese cars and to interview their owners. David Halberstam laid it all out in *The Reckoning*, a masterly study of Ford and Nissan through the early 1980s. Unfortunately, Halberstam published *The Reckoning* a bit too soon to account for changes that were affecting both automakers, portraying Ford as much weaker and Nissan a good deal stronger than they actually were. Still, the leaders of the Big Three dragged their feet, never quite believing that their industry, the bulwark of American prosperity, stood in imminent danger, ever hopeful that the U.S. government ultimately would rescue them from the encroaching Japanese tide.

By the time Lee Iacocca proudly repaid the government-guaranteed loan in 1983, Chrysler appeared to all the world to have solved its competitive riddles. Americans were buying cars again. The K car was a smash hit and the minivan was about to become one. The company hardly knew what to do with all its profits. In fact, the worst was yet to come: Japan's automakers were designing bigger and more luxurious vehicles than ever before, and they were planning to open factories in the United States. Iacocca, meanwhile, became dazzled by the power, perks, and celebrity of his office. As Chrysler's vehicles fell behind global competitors, he was busy giving speeches, exploring global tie-ups with other automakers and investing in nonautomotive businesses. Like the captain on the bridge of the *Titanic,* he apparently didn't see the danger that lurked just beneath the surface and didn't understand that his ship was far more vulnerable than he imagined.

By chronicling the activities of Lee Iacocca, Chrysler, and the Japanese automakers during the period following Chrysler's near-bankruptcy, I have tried to provide an understanding of the pervasive transformation affecting the entire U.S. auto industry and, indeed, all of manufacturing, from the executive suite to the factory floor. In order to explain how corporations are governed and how they arrive at decisions critical to their future — and to provide a dramatic framework — I have devoted special attention to Iacocca's on-again, off-again retirement and to the process of selecting his successor at Chrysler, as well as a brief look at his alliance with billionaire investor Kirk Kerkovian. In some ways this book is meant to elaborate, revise, and refine the Iacocca myth as shaped, for the most part, by his autobiography. The story of Iacocca, Chrysler, and Japanese competition concerns the evolution of management, technology, and marketing. It is also a story about people, particularly about behavior and politics at the level of chief executive.

This book is the culmination of ten years of reporting on the auto industry for the *Wall Street Journal,* the *New York Times,* the *Detroit*

Free Press, and as an author. This was a period that began during an economic recovery, deteriorated into the worst recession that Detroit has ever experienced, and ended during another recovery. One of the first things that a Chrysler public relations man tried to teach me upon my arrival was that things are never as good as they seem during a recovery, nor as gloomy as they seem during a recession. He was right. The auto industry is now selling cars as fast as it can build them, which casts a glow of optimism on nearly everything the automakers do or say. The next recession, whenever it occurs, will provide a truer test of how well Chrysler and the rest of the automobile industry have learned the painful lessons of the past decade.

SUNSET

THE DIRECTORS had been gathering courage for two years. Finally, in December of 1991, they decided to act. "Lee," one of the senior directors asked him, "would you mind stepping outside?"

Now he stood outside the big boardroom, staring from the lobby window of the twenty-fifth floor, the top floor of the American Center building. Below, he could see the traffic on Interstate 696 in Southfield, a city on Detroit's northwest border. The sleek building was the former home of American Motors, which Chrysler had absorbed four years earlier. To the northwest, a small forest was visible, and beyond it homes in Farmington Hills, another of the Motor City's bedroom communities.

Inside, the board was deliberating—minus its chairman. A few weeks earlier a Chrysler board meeting without Lee Iacocca would have been unthinkable.

Lee Iacocca attempted to subdue his emotions but the seething resentment and frustration were difficult to mask. Chrysler's nonemployee directors decided they had been patient too long. No one was indispensable. They had decided to pry Chrysler from his grip; they were going to find a successor.

In truth, the directors' decision to act was overdue. For twelve years, whatever Lee Iacocca said was law at Chrysler, from the line speed of the factories to the scripts of the television commercials touting LeBarons, New Yorkers, and minivans. Everything Chrysler

1

did was in some way a reflection of his taste, his enthusiasm, and—
after he became a big celebrity—his tendency to become distracted
from the basics of automaking. Twelve years was a long time for one
man to hold ultimate authority over a huge automaker, particularly
when the rules of manufacturing were changing so pervasively, favoring consensus-building skills over authoritarianism. No man was
able to be everywhere at once, to understand all the complexities of
technology and design. Yet an issue as seemingly trivial as whether to
add a chrome hood ornament to the Chrysler LeBaron was decided—
when Iacocca wanted it so—with a wave of his Montecristo cigar.
The board rarely felt inclined to challenge Iacocca's ideas. The directors paid him more or less what he requested. They approved the
acquisitions he desired. But suddenly, when it came to the most important career move of his life—the decision when and how to finally
hang up his spurs—the board had plucked the power from him.

For the past two years he had delayed, changed his mind, waffled.
Beginning just prior to his sixty-fifth birthday, on October 15, 1989,
everyone wanted to know when the illustrious chairman of the board,
best-selling author, and self-styled hero of the business world was
stepping down from his storied pedestal as the auto industry's most
colorful, most watched, most talked about figure since Alfred Sloan.
Who would replace him? Who could fill his shoes? And most importantly: Who could save Chrysler from impending disaster?

Outwardly Iacocca insisted that he wanted to leave. He just
wanted Chrysler to return to the kind of health that would allow him
to leave gracefully, a winner. At some point every chief executive
faced his dilemma. He was torn. He felt weary. His mother and daughters encouraged him to slow down. At times he felt eager to climb the
mountain one more time, lead his company once again to the promised land. At the same time, reasonable voices were telling him it was
time to step aside. When he heard these voices he became angry; he
couldn't bear the thought of relinquishing his spot to someone else.
Whatever happened, he wanted the decision to be his.

He thought the issue of succession had been settled. Then, one
by one, the candidates had grown discouraged and quit or had been
disqualified. Jerry Greenwald, after waiting in the wings for almost a

decade to take over as chairman, quit to lead a buyout attempt of United Airlines. Roger Penske, the dashing owner of an Indy 500 racing team, was an eyelash away from agreeing to lead Chrysler. Then Penske pulled out at the last minute. The directors thought Jerry York, the new finance man, was too young. Steve Miller, Iacocca's longtime in-house dealmaker, had been a remote possibility. But now Miller was out too, due in part to an indiscretion that had infuriated Iacocca.

Bob Lutz, the man steering the new generation of Chrysler cars and trucks that were about to appear, had stuck around. He wanted the chairman's job badly. To the outside world, Lutz seemed a natural. He was smart, experienced, talented: he could lead. Lots of people inside Chrysler were eager for him to be named chairman, and a few of the directors were quite impressed with his leadership potential. But Iacocca would be damned before he agreed to let Lutz take over as chairman. He had reasons, good ones. No, Lutz was not going to be the man either. Iacocca seemed pretty sure that was to be the case — at least until the afternoon when the board asked him to step outside while they deliberated the issue of executive succession.

Iacocca had watched the board grow more independent a few months earlier, after he had been quoted in the *Detroit News* as implying that he was thinking of remaining at Chrysler indefinitely. For the past two years a small group of directors had been gently urging him to create a succession plan, warning him what might happen if it dragged on too long. He'd enjoyed a warm and friendly relationship with Paul Sticht, the retired chief executive of RJ Reynolds, with Bob Lanigan, the retired chief executive of Owens-Illinois, and a few others. But lately, they — like many other of Chrysler's corporate directors — had developed a view fiercely independent of the chief executive's. You've done a great job, Lee. We have nothing against you, they said. Do it on your own schedule, but take care of it. A public corporation simply can't operate without a schedule for identifying and promoting top executive talent, *including the chief executive.* As directors, we're responsible for smooth transition, they had told him. Our lenders, employees, and shareholders want it that way, and they are entitled to managerial depth.

On August 27, 1991, reporters had staked out Iacocca during a ceremonial appearance at a new Dodge dealership in Rochester Hills, Michigan. Normally he felt that to appear publicly at a dealership opening was beneath him, but in this instance the owner was a friend. Chrysler's financial losses and deteriorating credit ratings had been a subject of intense press attention for months. The reporters who showed up had come to see him, not the new dealership. They waited near his trailer, wanting to find out about his retirement plans. "I'm really sixty-four," he told reporters. "Print that. I've got a couple years to go." (In fact he would be sixty-seven in two months.) Directors weren't even going to begin thinking about his successor until the following year, he said. That bizarre statement, meant to taunt and confuse the reporters was plastered in the newspaper the following day. When directors read his words, they were determined to take control of succession, which meant taking the process away from Lee. Indeed, directors had been deliberating for the past year and had been urging him to propose a candidate. They certainly didn't intend to wait until next year, 1992, to start thinking about a successor. And they would not allow anyone—not even Lee Iacocca—to preempt their duties as directors.

Iacocca stewed as he waited near the boardroom, gazing at the suburban Michigan countryside. His mind was working: as he stood there he must have concluded that the game wasn't up, not by a long-shot. The outside directors had the power to exclude Iacocca from their discussions, but he knew they didn't possess the clout to force him out of Chrysler without a fight. He had arrows left in his quiver. He had fought in enough corporate wars—and learned enough about back-alley fighting from Henry Ford II—to know that several options were still open to him. He knew a way, the perfect strategem, to hang on to his power and his pedestal—and satisfy the Chrysler board as well.

EIGHTY YEARS
OF SUPREMACY

THE POWERFUL, sophisticated industrial empire that Lee Iacocca stood astride in the early 1990s might have looked surprisingly formidable to those who remembered the Chrysler Corporation of twelve years earlier but not to those who knew well its past. Like the history of many automakers, Chrysler's was a weaving, erratic affair, sometimes brilliant, sometimes misguided, marked by the parade of pioneers, plungers, and industrialists who ran it.

Chrysler was founded in the first part of the century, a period when hundreds of automakers eked out a living before failing or being merged into larger enterprises. Chrysler survived by distinguishing itself for excellent engineering, the legacy of Walter Chrysler, who gave the company its name. But for most of its history Chrysler was a marginal enterprise, floating along in strong economies, scrapping for life during recessions — an industrial bungee jumper hurtling toward disaster one moment, snapping back to safety the next. The directors of Chrysler who were deliberating about replacing Iacocca believed that disaster, and possibly oblivion, awaited them if they didn't act swiftly and decisively.

It was remarkable how automaking, which had changed much in a relatively short period, remained in many ways the same. In the late nineteenth and early twentieth centuries the founders of the American automobile business survived by dint of innovation, earning and losing great sums of money and gouging a deep impression in the industrial, social, and economic landscape. At the height of their success, many automakers—swollen by ego or blind to changing conditions—stumbled and then recovered. A pattern emerged: auto companies survived and grew insofar as they were able to please car buyers during periods of prosperity.

Invariably, a period of success invited a period of neglect. Automakers used their earnings to hire more employees, raise salaries, redecorate offices, and generally increase costs. If pockets of waste and inefficiency were created, this was viewed as a trivial issue because costs were insignificant next to the vast profits that would be made during times of plenty. Of course, when recession suddenly hit or the public turned its nose up at a new model, automakers rarely cut costs fast enough to prevent ruinous losses.

Billy Durant, the entrepreneur who created the General Motors Corporation from a collection of individual carmakers in the early 1900s, was an impulsive dreamer whose business acumen never matched his audacious vision. Bankers bounced him from his own company twice, replacing him finally with Alfred Sloan, whose legendary ideas about management are studied to this day. Henry Ford, a contemporary of Durant, showed the world how to organize mass production but nearly wrecked the Ford Motor Company through disorganization and a financial backwardness borne of paranoid suspicions.

Walter Percy Chrysler was one of the very few great automotive men whose careers never suffered a serious reverse. Born in Wamego, Kansas, in 1875, Chrysler showed mechanical aptitude as a youngster and longed to work for a railroad, as his father did. He signed with the Union Pacific as a machinist's apprentice, studied mechanical engineering by mail, and became a master mechanic. In his spare time he played the tuba.

In 1908, the year Henry Ford introduced the Model T, the Chicago Great Western Railway hired Chrysler as "superintendent of motive power," the youngest man ever to hold the job. On a visit to one of the first Chicago Auto Shows, Chrysler was smitten by a five-thousand-dollar Locomobile, an ivory and red number with headlamps fueled with oil. He didn't know how to drive a car but burned with curiosity to understand the machine. He bought it with his savings and $4,300 borrowed from the Continental National Bank of Chicago. Endlessly, he took it apart and reassembled it. "I did not simply want a car to ride in, I wanted the machine so I could learn all about it," he wrote. "These self-propelled vehicles were by all odds the most astonishing machines that had ever been offered to men."

Chrysler's next job took him to the American Locomotive Company in Pittsburgh (where he bought his second automobile). His achievements there building steam locomotives drew the attention of James J. Storrow, a director of American Locomotive. Storrow, who was the head of a Boston investment bank, was also in charge of a group trying to reorganize General Motors. Chrysler's enthusiasm for cars must have been evident since Storrow was able to recruit him as manager of GM's Buick factory in Flint, Michigan, at $6,000 a year, half his previous salary.

Using the concept of moving assembly lines developed at Ford and at Cadillac, Chrysler was able to boost production over three years from a near-standstill to 140,000 units annually. By 1914, after two years on the job, Chrysler felt confident enough to demand a raise to $25,000. He demanded $50,000 the following year and was granted it. By 1919, Chrysler was earning $500,000 a year as head of the Buick division. But he also clashed with Durant over the latter's questionable business judgment. In March of 1920, at the age of forty-five, he quit General Motors.

"I remember the day," Sloan would later recall. "He banged the door on the way out, and out of that bang came eventually the Chrysler Corporation."

In less than a year, bankers holding $50 million of loans to the Willys-Overland company asked Chrysler to try and save the automaker. The bankers were willing to pay him $1 million annually. He took the job, he later wrote, because his wife was weary of his sitting around the house smoking cigars with friends.

Willys-Overland, one of the many struggling automakers, had prospered from the boom of World War I. With the flood of cash it bought an airplane company, a harvester company, too many parts and too much equipment. In Chrysler's opinion, "all that was bad in the Willys-Overland Corporation was due, really, to lack of competition, to the wartime boom and its easy money." The seeds of an endemic ailment had been sown: prosperity had produced car buyers in droves, which led to managerial slackness and diversification into fields that automakers didn't understand.

Chrysler's first step was to reduce Willys-Overland's debt. John N. Willys, however, opposed the new model that Chrysler was developing with the help of three young engineers — Fred Zeder, formerly of Studebaker, Owen Skelton, and Carl Breer. When Willys's opposition stiffened into resistance, Chrysler tried to take over the company; later he tried to buy it at auction. Both tactics failed, so in 1922 Chrysler decamped, taking Zeder, Skelton, and Breer with him from the Willys-Overland offices in Elizabeth, New Jersey, to a vacant space in Newark to work on the design of a new car.

Simultaneously, Chrysler was assisting bankers in untangling the finances and operations of the Maxwell Motor Company in Detroit. Maxwell was building cars at a factory on Jefferson Avenue. They were of poor quality and, as a result, not selling. Under Chrysler's direction, Maxwell declared a financial loss on the cars and sold them at deep discount. At the same time, Chrysler decided that Maxwell's ultimate salvation lay in a new model, the very one he was developing.

In the fall of 1922, Walter P. Chrysler transferred his trio of engineers from Newark to Maxwell's Jefferson Avenue factory. He enthused over their design of a new high-compression engine that generated more power with less noise and vibration than the typical six-cylinder engines of the day. By the fall of 1924 Maxwell was nearly out of money. A prototype of the new model, powered by the new engine, went on display during the New York Auto Show in December. Chrysler had to lease the lobby of the Commodore Hotel to display his car since the trade show refused cars that weren't already in production.

The prototype, the first to bear the Chrysler badge, was the hit of the show. Its low-slung posture attracted praise, and it featured, in addition to its engine, several technical advances, such as a replaceable oil filter and four-wheel hydraulic brakes. Most importantly, the prototype impressed the bankers, who were willing to lend the

$5 million needed to start production. On the strength of this loan, Chrysler incorporated the Chrysler Motor Corporation, which took over Maxwell Motors in early 1925 in an exchange of stock.

The phenomenal success of Chrysler's first car, the Chrysler Six, set the stage. First-year sales of 32,000 were the most ever for a new model, far outselling the existing Maxwell line. The Chrysler Four followed the Six, signifying early recognition of different groups of buyers with varying tastes and budgets. A year later, in 1926, the automaker introduced the six-cylinder Imperial 80, a luxury sedan featuring an instrument panel with ivory dials. The "80" signified the Imperial's top speed in miles per hour—at a moment when Ford's Model T, the most popular model of its day, reached only thirty-five miles per hour.

Within a few years of its founding, Chrysler Motor had gained a nationwide reputation for engineering excellence. Each year the automaker added more creature comforts, like adjustable seats, or safety features, like shatterproof glass. For the fashion-conscious, Chrysler offered wire wheels and painted monograms. Still, Chrysler was only the fifth-largest U.S. automaker in 1927, with sales of 192,000, dwarfed by the number-one Ford and number-two General Motors, which sold about one million cars each that year.

By sticking too long to his Tin Lizzie Model T, Henry Ford demonstrated for Chrysler and other automakers a clear and crucial mistake to be avoided. Ford thought customers wanted cheaper, plainer, more utilitarian transportation when quite the opposite was true. Many customers desired, and were willing to pay for, style and performance. General Motors's Sloan captured the trend with the slogan "a car for every purse and purpose."

The more subtle dimension of Ford's mistaken reliance on the utilitarian car was his single-minded executive obstinacy, which overshadowed valid ideas from colleagues and subordinates who better understood the evolving automotive culture. It was the classic trap for someone who was so often right he couldn't imagine being wrong. Ford's blunder let General Motors, under Sloan's brilliant leadership, gain an edge that has endured for more than half a century.

By the late 1920s, Walter Chrysler understood that his company must grow larger if it was to survive. The massive investment requirement for new models outstripped the resources of tinkerer-entrepreneurs. Chrysler had to buy parts-making plants so he wouldn't be at the mercy of outside vendors. He needed more assembly-plant capacity in order to build several models to serve the proliferating tastes and budgets of consumers. The Dodge Brothers of Detroit owned the parts-making and assembly capacity Chrysler needed. John and Horace Dodge had both died during the influenza epidemic of 1920. In 1925 the company was sold to Dillon, Read and Company, investment bankers, who were struggling with the operations. With his eye on the "splendid plants of the Dodge Brothers," Chrysler bought Dodge in 1928 from Dillon, Read for $100 million in Chrysler stock and the assumption of $70 million of Dodge debt. Overnight Chrysler was the third-largest automaker in America, with 12,000 dealers and the capacity to build up to one million cars annually.

Chrysler added the Plymouth brand, a down-scale car meant to compete with Ford's utilitarian offerings, and the DeSoto brand, with which he hoped to seize the middle market between the Plymouth and the upscale Chrysler brand.

Having seen Maxwell's quick, poorly planned expansion dissolve into failure, Chrysler felt compelled to prove that it wouldn't happen to him. "If we grow, we will grow according to our financial, marketing and manufacturing capacities — not beyond them, in a straining after domination." Because Chrysler consciously avoided overextending, the company sailed through the Great Depression, largely on sales of the $600 Plymouth to many who were strapped for cash. Indeed, the Plymouth factory in Detroit had the biggest daily production capacity of any single car under one roof. Rather than recruit a separate network of Plymouth dealers, Chrysler sold the car through existing Chrysler, Dodge, and DeSoto dealers, thus keeping many dealers afloat when demand for expensive cars was low.

Walter Chrysler preached the importance of research and boasted that he didn't cut a penny during the Depression. In 1934 the automaker brought out the Airflow, a bold project that introduced several new design features and technologies, such as the single-piece curved windshield, headlights built into the fender, and the automatic overdrive transmission. The Airflow, while praised by many, taught Chrysler a lesson about the mysteries of customer taste, for the model was soundly rejected and dropped from production in 1937.

As automakers introduced more optional features and colors, manufacturing complexity increased exponentially. If automakers built what they guessed would be the most popular combinations, they risked being stuck with thousands of unsold cars. Responding only to dealer orders, however, required tremendous coordination and communication among dealers, assembly plants, and parts suppliers. Nevertheless, by 1935 Chrysler had imposed a strict policy of building cars only to dealer order, and a system of sending thirty-five copies of every order for the planning of each day's production. Each car was shipped seven days after the order was received.

Although Chrysler's disciplined ordering policy minimized wasted material and inventory, it aggravated the hassles associated with seasonal fluctuations in sales and production. People loved to buy cars in the spring and tended to lose interest in shopping during the summer. Slowing production after the spring meant laying off workers; labor was a controllable cost, and prior to 1937 the United Auto Workers didn't yet exist to guarantee annual wages and benefits.

The strain of production, periodic layoffs, and the increasing complexity of new machinery imposed burdens on workers that Alfred Sloan, Henry Ford, and Walter Chrysler only dimly appreciated. Although autoworkers' pay was good for the 1930s, the car plants were hot, dangerous, exhausting places to work, ruled in many cases by overbearing bosses who intimidated subordinates and weren't above using physical violence to maintain production schedules.

Chrysler's clashes with factory workers were ugly, though not quite as severe as those at Ford and General Motors. On a political level, Chrysler lobbied Congress against legislation that provided for collective bargaining, but it also paid bonuses, raised wages, and tried conciliatory gestures like posting World Series scores in the plants. When the United Auto Workers Union won the right to represent General Motors workers after the sit-down strike of February

1937, less than a month passed before a similar shutdown hit Chrysler's assembly plants. After a second massive strike in the fall, the union won full recognition to bargain on behalf of autoworkers.

By 1935, Walter Chrysler had lost his zeal for automaking and was ready for a successor. Kaufman Thuma Keller (known to everyone as K.T.), a manufacturing specialist, had been recruited from General Motors in 1926. A master mechanic, his first big job was consolidating the Dodge Brothers and Chrysler plants and making room for DeSoto production. Keller was a classic micromanager. From time to time he personally supervised the installation of a single machine. The smallest details fascinated him. He insisted, for example, that initials rather than first names appear in the company directory.

Without the auto companies, the Allies' victory in World War II would have been far more difficult. Automotive assembly lines produced airplanes, tanks, jeeps, and weaponry. Within eight months of Pearl Harbor a new Chrysler tank plant stood in Warren, Michigan. By war's end the plant had produced 25,000 tanks. Chrysler also built 438,000 trucks, 18,000 B-29 aircraft engines, 60,000 antiaircraft guns, 7,888 marine tractors, 29,000 marine engines, 101,000 incendiary bombs, 100 miles of submarine netting and countless other tools for the military.

President Truman decorated Keller for Chrysler's achievements in aiding the war effort. But under his leadership the automaker's management methods slackened. Like his chief financial officer, B. E. Hutchinson, Keller believed that during periods of strong sales and profits, Chrysler was entitled to increase salaries and dividends. The increased outlays for salaries and dividends were achieved in part by holding down spending on factories and modernization. Keller also exercised his predecessor's prerogative to dabble in styling and to impose his personal taste on new models.

By this time, however, Chrysler had grown into an expansive, mass enterprise — beyond the scope of a chief executive to decide relatively minor issues. The world had never seen private enterprises the size of Chrysler, Ford, or General Motors, and there were no textbooks on how to run them. The influence of entrepreneurs was giving way to the influence of managers. Keller didn't rise to the top because he demonstrated special ability to lead an unusually large group of managers or complex, far-flung operations. Through forcefulness and dynamism he won the ear and the personal favor of Walter Chrysler over Hutchinson and Zeder — and he, for better or worse, was chosen chief executive.

The Ford Motor Company had been even less successful in finding a management system. Harry Bennett, a ruthless thug, took advantage of Henry Ford's failing mental capacity to seize decision-making powers, nearly running the automaker into the ground. The last-minute intervention of young Henry Ford II, who was released from naval service during the war to rescue his family's company, prevented Bennett's unsavory influence from destroying the company.

Only General Motors, under the guidance of Sloan, devised anything resembling a rational way to manage its enormous scale. Sloan devised "yardsticks" to measure the performance of General Motors executives. He made sure that talented candidates gained experience in several jobs and prohibited personal considerations from coloring personnel decisions. Sloan's methods weren't perfect but they were far superior to the chaotic executive politics at Ford and Chrysler.

After the war, two key events shaped Chrysler's destiny. The first, undoubtedly, was the economic boom sparked by the Allied victory. Along with new housing, Americans had worked up a terrific appetite for new cars. Although in retrospect, Sloan was correct in his belief that customers would increasingly view cars as expressions of fashion and style, postwar demand was so strong that factories could not be converted back to domestic car production fast enough to keep pace. Customers were buying almost anything that rolled, stylish or not.

Keller held views opposite to those of Sloan. Cars should be practical, he believed. Attempting to satisfy fashion was dangerous. Perhaps the memory of the Airflow debacle lingered. While Ford and General Motors were designing lower, rounder, sleeker models, Chrysler's models were tall and boxy. Keller had countermanded his stylists, adding several inches of height with the explanation that "men and ladies are in the habit of getting behind the wheel, or on the back seat, wearing hats." When reporters questioned this insight, he replied, "Chrysler builds cars to sit in, not piss over." His cars were selling well, so who was to say he was wrong?

Keller's management style also contrasted sharply with that of Sloan. Sloan decentralized GM's operations, encouraging executives to think independently and be judged by the outcome. Keller, on the other hand, insisted on supervising every major department. Likewise, each Chrysler department was a fiefdom unto itself, headed by a powerful vice president who decided substantive issues, and those as minor as who received favored parking spots. And so it proceeded, from the top down to the very lowest rung — level upon level of workers with limited authority, looking upward for guidance, rarely questioning a judgment from above.

The second great event affecting Chrysler was the revival of Ford. Twenty-eight-year-old Henry Ford II, after throwing out Bennett and his henchmen, recruited Ernest R. Breech, a former General Motors executive to help rebuild his family's company. Ford also hired ten former air force officers who had specialized in statistics and logistics during the war — a group that would become known as the "Whiz Kids" — to use their expertise to modernize Ford's operations. Just as important, Ford in 1949 brought out a new sedan — lower, lighter, and without a running board — that appealed to the public's taste far better than Chrysler's outmoded Plymouth.

The 1949 models were crucially important, for they were the first new models since the war. Keller consolidated his control by taking over as chairman in 1949, a post left vacant since Chrysler's death in 1940, but at the same time he failed to read the signals that customers were sending. The year of reckoning was 1953, when Chrysler forever lost its hold on second place to Ford. Smaller automakers like Crosley, Kaiser-Frazer, and Willys-Overland were getting hurt as well. The Hudson Motor Car Company and Nash-Kelvinator Corporation merged in 1954 to become the American Motors Corporation. Chrysler had made its reputation on engineering innovation. A Plymouth, therefore, normally commanded a higher price than similar Ford and Chevrolet models. But with many of the period's major innovations, such as the V-8 engine and automatic transmission, already introduced and adopted generally, automaking turned into a contest of superior marketing, and the basis for Chrysler's price premium vanished.

A car's appearance became perhaps the single most important factor to consumers of the 1950s. Changes in the sheet-metal exterior, which previously had come every four or five years, became an annual event. Changing exteriors was expensive. Every metal stamping that was welded into the body had to be changed, necessitating new dies and reconfigured body shops. Retooling costs might run as high as $100 million; so automakers had to see profits from a new model immediately, since the body had to be restyled for the following year.

Chrysler's achievements as a military contractor in World War II gained it a foothold in space technology. President Truman appointed Keller to organize the Pentagon's guided missile program. In 1952 a few dozen Chrysler engineers were assigned to work on the Redstone missile in Huntsville, Alabama. With Keller's time thus divided between cars and rockets, Chrysler's car operations fell increasingly under the influence of Lester L. "Tex" Colbert, a protégé of Walter Chrysler who was named president in 1950. Colbert, a Harvard-trained lawyer, had handled labor negotiations in 1937 and served as president of Dodge after the war.

Colbert came over to Chrysler with the toughminded insight that the company's postwar sales stemmed more from a boom economy than from its prowess as an automaker. One of his first acts was to hire McKinsey and Company to make a thorough, independent view of Chrysler's strategy. For Chrysler, hiring an outside management consultant constituted a crucial admission that all knowledge did not reside within its walls. McKinsey suggested, among other things, that Chrysler expand overseas and decentralize its decision-making. In short, McKinsey advised Chrysler to become more like General Motors and, to a lesser extent, Ford.

McKinsey also advised a complete redesign of the cars' exteriors. Colbert hired Virgil Exner, a Studebaker stylist, and put him to work on the 1955 models. Exner stretched the cars, gave them jazzy two-tone and three-tone color combinations, and in 1957 endowed them with fins. Despite Chrysler's expenditure of nearly $1 billion to modernize plants, the aggressive, frequent styling changeovers came at the expense of manufacturing quality. Wind and water leaks were numerous. Cars rusted and parts fell off.

Colbert fared little better with decentralization. He learned what future generations of auto executives would discover: a corporation's managers can frustrate the plans of a leader without ever disobeying his commands. Subordinates go through the motions without enthusiasm, causing initiatives to fail. Colbert's underlings rejected decentralized power, and would always do so, it seemed, as long as they perceived that others controlled their fate. As long as the penalty for mistakes was high—loss of promotion or bonus—managers always chose to do nothing when possible. Mistrust and finger-pointing among managers was rife. A rigid hierarchy provided layers of protective insulation for the low-ranking. Disputes were pushed upward to the executives who were paid to battle one another.

On the international front, Colbert expressed interest in Rolls Royce and Volkswagen. He succeeded in buying a 25 percent stake in the French automaker Simca. But the prime assets to be acquired overseas already had been bought by GM and Ford.

The recession of 1958 exposed flaws in Chrysler's postwar strategy. Chrysler's customers, who were less affluent than those of its competitors, were hit harder than other segments of the population. As sales dropped, Chrysler scrambled to cut costs by closing plants and laying off workers. Production was reduced by about 50 percent, but fixed costs remained high. General Motors and Ford were hurt too, but only Chrysler recorded a loss in 1958. Its financial men scrambled to line up $150 million in credit from 100 banks, just in case.

Chrysler was caught in a squeeze between General Motors and Ford above, and the makers of small, inexpensive cars below. Despite the contempt of Detroit executives for small, low-priced cars, import sales had been creeping upward; combined with Studebaker and Rambler sales, they accounted for about a fifth of all car sales by 1959. General Motors, Ford, and Chrysler hastily began producing their own small cars in the early 1960s—the Corvair, Falcon, and

Valiant—but the profit margins on these cars were far slimmer than on large cars. The reason was simple: although it took just as much engineering, tooling, and labor, and almost as much material, to build a small car as a large one, the prices commanded by small cars were far less.

As automaking confronted upheaval in consumer taste, Colbert was, at best, struggling to manage the change. Colbert's protégé and designated successor, a man named William Newberg, found himself the target of controversy in 1960 before he even had a chance to properly take over. Newberg resigned sixty-four days after being named president. According to its 1961 proxy statement, Chrysler claimed that Newberg and his wife improperly owned stakes in two of Chrysler's part suppliers. The claim was settled.

George Love, a coal executive from Pittsburgh who had been recruited by Colbert as a director, believed Chrysler desperately needed a leader with a gift for financial analysis. After all, GM had a finance man, Fred Donner, at the helm. Ford's operations were ruled by finance men. Love became chairman and promoted the candidacy of Lynn Townsend as president. Townsend, a forty-two-year-old accounting specialist, joined Chrysler as comptroller in 1957. Though he sported sideburns, which was enough to make him moderately eccentric by Detroit standards, Townsend epitomized, in the slightly contemptuous argot of business, the bean-counter.

The auto industry's engineers and designers understood why strong financial management was necessary, particularly in mass enterprises, yet they feared and suspected "accounting types." They knew that "beanies" made decisions based on numbers, which were abstract, rather than on concrete physical principles. They resented being ordered to scrimp and save on material by bookkeepers, who were ignorant of how that spoiled quality.

Townsend, though without vast automotive knowledge, was a bright man with a broad view. Armed each morning with three-by-five cards detailing sales and production, he did indeed understand costs and how to cut them. Faced with fixed costs and declining sales, he cut his white-collar staff of 28,000 by 25 percent in 1961, in one case replacing 700 clerks with an IBM mainframe. He consolidated operations and sold off company airplanes.

At the same time, Townsend realized that a corporation's image was so important as to almost defy a price tag. To revive faith in the Chrysler name, badly tarnished by the disastrously poor quality of their 1957 models, he instituted a five-year, 50,000-mile warranty, far beyond the industry's norm of guaranteeing cars for one year or 12,000 miles. Townsend spent $50 million with a consulting firm to create the "Pentastar" logo and conduct an image campaign.

Within months of taking office, Townsend jumped into the engineering of current models and the planning of the 1963 cars. Any manager who couldn't improve quality while maintaining production was shown the door. Almost immediately the dealers reported improved quality. He recruited a top sales executive from American Motors and a top designer from Ford. Robust economic growth brought car buyers back to the showrooms, and the company's balance sheet showed strong revenue and profit. By the end of 1962, Townsend was toasted by financial writers and celebrated on the cover of *Time* as the architect of Chrysler's comeback.

The 1960s, a period of rapid economic expansion for the nation, greatly brightened Chrysler's outlook. Townsend was investing money in new plants as fast as the cash came in. He also raised $263 million in an offering of common stock, the company's first since 1928. In 1963, Chrysler's capital spending was projected at $700 million over ten years; in 1964, he raised it to $1 billion over four years; and the following year the amount was increased to $1.7 billion. New assembly plants were built in Belvidere, Illinois, and St. Louis, as well as a stamping plant in Sterling Heights, Michigan, and new foundries in Detroit and Kokomo, Indiana.

Chrysler also expanded its international portfolio. It increased its stake in Simca to 64 percent. In 1965 Chrysler bought a large stake in Rootes, a British automaker that built the Hillman and Sunbeam brand cars. The next acquisition was a Spanish truckmaker, Barreiros, followed by acquisitions in Latin America. Toward the end of the decade, Chrysler operated plants in eighteen countries.

Improved access to credit and equity markets drew Chrysler to real estate. The automaker invested in luxury housing, student dorms, and the Big Sky ski resort in Montana. The real estate subsidiary's main purpose, however, was to buy and to finance land and showrooms for new dealerships. Townsend grasped the importance of securing prime, high-traffic locations for retailers to sell cars.

Government contracting through the space program added revenue, as did Chrysler's tank business, which won the contract for the army's M-60 main battle tank. Chrysler diversified into other fields as well, including chemicals, air conditioning, and pleasure boats. These operations consumed a good deal of executive attention but remained proportionately tiny next to the car business.

In 1964, Chrysler finally opened its own credit subsidiary, a field that General Motors developed in the 1930s and Ford entered in 1950. Prior to automakers' "captive finance" companies, car buyers depended on banks, credit companies, and credit unions for new-car loans. When credit was tight, banks turned fickle, causing dealers to lose sales. As long as credit subsidiaries like General Motors Acceptance Corporation (GMAC) lent prudently, they could pocket the profit the banks had been making on loans while assuring access to credit for the purchase of their own cars. Finance subsidiaries also helped dealers pay for the cars they bought from the factory.

Carried along on the tide of a surging economy, Townsend could do no wrong. Little by little, always carefully tracking the numbers, he increased sales and production volume. Some of the increase was due to general growth in total automotive sales, though Chrysler was commanding a bigger share as well, 16 percent of the market in 1968, as compared to 11.3 percent in 1959. In 1965 the Chrysler board named him chief executive, and in 1967 he took over as chairman from the retiring George Love.

"If the Chrysler Corporation's silver cloud has a murky lining, it is a hard one to detect," observed a writer in the *New York Times* of August 9, 1968. "Almost every move made by Lynn Townsend and his management since 1961 seems to have been the correct one."

The automakers, especially Chrysler, were living in a fool's paradise. The insatiable appetite of car buyers of the 1960s blinded

Townsend, the union, and many others to economic traps about to snap shut. Easy credit and rising wages gave millions of Americans the ability to buy new cars every year or so. For the automakers, rising costs didn't matter since they were transferred to customers as higher prices. The only statistic that mattered was production. If assembly lines moved, cars departed for the dealerships, where customers awaited.

In a market where demand appeared to be bottomless, quality often became secondary. Manufacturing defects were repaired at the end of the assembly line. The only unforgivable sin was failing to meet a daily production quota. Townsend and his lieutenants didn't attempt to forecast how many customers wanted Chrysler cars. They worked backward from quarterly earnings. The value of executive stock options depended on the price of the stock. In large measure, stock price depended on earnings. Thus, every dollar of profit required a specific number of cars produced. If dealer orders didn't match production targets, the cars would be built anyway and executives assigned the excess to a "sales bank." Sales managers cajoled and pleaded with dealers to buy more cars, offering discounts and prizes, such as trips to exotic locations, as incentives. Thousands of "sales bank" cars were shipped to fairgrounds, railroad yards, and other storage areas. Weeds grew into their wheelwells. Repairmen roamed the muddy fields, inflating tires, recharging batteries, and repairing damage.

In their lust to ship vehicles Townsend and other Chrysler executives virtually abandoned the study of consumer preference. Were customers buying convertibles? How much were they willing to pay for air conditioning? Were smaller cars gaining in popularity? Sales reports were routinely inflated and, therefore, yielded little meaningful information about the market. Chrysler failed to mount a credible sporty car to challenge Ford's Mustang. The Chrysler Imperial ran a poor third to Cadillac and Lincoln in the luxury segment. The Dodge Dart and Plymouth Valiant sold well, though economy cars produced meager profit.

At one point in 1966, Chrysler's sales bank contained more than 60,000 unsold cars. The first time Chrysler needed to clear massive inventory, it did so by extending favorable credit terms. Later on, dealers were pressured to accept sales-bank cars if they wanted access to the hot-selling models. Now and then Chrysler threw in cash, trips, furniture—whatever it took—to persuade dealers to accept cars. Dealers didn't like the system, sales executives hated the pressure, and factory chiefs worried when hastily assembled cars were shipped in the closing hours of a fiscal quarter—all in the name of meeting production quotas designed to keep profit and stock price afloat.

The merry-go-round screeched to a halt in late 1968. Car buyers abruptly turned their backs on Chrysler's 1969 models. Imports like the VW Beetle, which had been on the wane for a few years, were making a comeback. Several years of pushing production volume while ignoring market trends spelled disaster for large-body cars: the Dodge Monaco and Polara, the Chrysler 300 and Imperial, the Plymouth Gran Fury. Chrysler posted a profit for 1968; but by February of 1969, the unsold inventory of cars had grown to more than 408,000!

Needing a scapegoat for the unsold cars, Townsend turned to Virgil Boyd, Chrysler's president. Boyd was forced to retire, while Townsend swung into action, firing 12,000 workers, trimming the dividend by 70 percent, and cutting costs until the failed models could be restyled. As Boyd's replacement he named John J. Riccardo, like himself a former accountant. With import sales on the rise, Chrysler needed a small-car strategy. Ford had the Maverick, and General Motors the Vega. Townsend, convinced that the number of Americans willing to tolerate a small car didn't justify the cost of development, unsuccessfully tried to plug the gap with Simcas from France and Hillmans from Great Britain.

By 1970, America's decade of growth and optimism was replaced by a sense of gloom and limitation. The Vietnam War was being bungled militarily and politically. Unemployment and inflation were creeping upward. The Middle Eastern oil-producing nations were forcing price increases on the oil companies. Chrysler suspended construction of factories in Kokomo, Indiana, and New Stanton, Pennsylvania, and was forced to borrow $200 million to tide it over. Cash was very tight. Under Townsend's leadership, Chrysler was

transformed from a virtually debt-free automaker to one with almost
$800 million in obligations, much more highly leveraged than Ford
or GM.

Indeed, when Chrysler concluded an agreement to buy 35 percent
of Japan's Mitsubishi Motors Corporation in 1971 — in order to import
Mitsubishi subcompacts under the Dodge Colt name — Townsend
discovered that he didn't have enough cash to sign the agreement.
The problem was rectified by buying the stake in two stages — 15
percent now and 20 percent later. But the sloppy negotiations were a
sign to the Japanese that Chrysler was not nearly as sophisticated in
its business dealings, or as financially sound, as they first imagined.

Chrysler was skating on even thinner ice than it knew, a fact
that first became publicly evident during the Penn Central collapse.
Hours after the railway filed for bankruptcy on June 22, 1970, holders
of $82 million in Penn Central commercial paper realized they had
lost their money. Commercial paper, short-term unsecured loans that
are traded publicly, was also sold by automotive finance companies
to raise money for car loans. Chrysler Finance had about $1.5 billion
worth of commercial paper outstanding, and investors were suddenly
afraid that Chrysler, which lost $27.4 million in the first quarter of
1970, might be the next Penn Central. The subsequent sell-off of
Chrysler commercial paper left the company with no way to raise
money for financing cars. The Federal Reserve turned down Chrysler's plea to ease the money supply as a means of calming the panic
in financial markets. The Nixon administration had been in no mood
to bail out Penn Central, and the same could be said of the politically
independent body that controlled the nation's money supply. Arthur
Burns, chairman of the Fed, accused Chrysler of trying to "turn this
country into a Latin American banana republic."

Townsend and John McGillicuddy, vice chairman of Manufacturers Hanover, Chrysler's main bank lender, crisscrossed the country,
entreating other bankers to keep the faith. The banks extended a
backup credit line of $180 million for Chrysler Financial, which
helped it to regain access to the commercial paper market. The automaker agreed to contribute $100 million to its finance subsidiary, and

the banks bought $150 million of Chrysler financial loans, thereby reducing the subsidiary's exposure. Calm was restored—barely.

The U.S. auto industry's darkest nightmare unfolded in 1970, a watershed year for new federal regulation. Ralph Nader had started wheels turning in Washington in the mid-1960s with his scathing indictment of the General Motors Corporation's negligence in the field of automotive safety. The National Traffic and Motor Vehicle Safety Act of 1966 might fairly have listed Nader as a cosponsor. By 1970, all sorts of new safety regulations took effect, requiring redesign and retesting of hundreds of thousands of parts that could cause death or injuries in the event of accident. Only vigorous lawsuits by Chrysler and others prevented mandatory installation of airbags by 1975.

The Clean Air Act amendments of 1970 bespoke popular worry over the automobile's contribution to air pollution. The law required automakers to reduce tailpipe emissions significantly by 1975. Chrysler wanted the law to be worded in a manner consistent with its approach to clean-air research. Chrysler's lean-burn combustion technology would have created cleaner engine emissions, while General Motors favored adding a "catalytic converter" that removed pollutants from emissions after they left the engine. In the end, GM's lobbyists won the day, ensuring that its catalytic converter would become the industry standard.

Either way, automakers faced billions in new expenditures for research, testing, design changes, and tooling expenses. Chrysler would have been severely disadvantaged even if its finances weren't in disarray, because the high fixed cost of safety and air pollution research and parts manufacture had to be spread over a smaller number of vehicles sold than at Ford or General Motors. Chrysler was slow to grasp just how much harder regulation was affecting it than its competitors, because the company's accounting methods didn't break out costs incurred to comply with new laws—such as an engine part whose sole purpose was to reduce emissions. If Townsend understood Chrysler's disproportionate burden he didn't show it.

By 1972, car buying and the economy were on an upswing once more. The cost-cutting of Townsend and Riccardo was working, at least temporarily. Profit reached $220 million, a big turnaround from the losses of 1970, prompting Riccardo to gloat: "The day this company turned around was the day we decided not to build that subcompact. Chrysler no longer operates by knee-jerk reaction. It does not imitate everything competitors do." Townsend raised the dividend and resumed plant construction, including the facility at New Stanton. Chrysler planned to borrow another $150 million of long-term debt.

The Yom Kippur War broke out in October of 1973, just as Chrysler's new large cars, representing an investment of $450 million, were to be introduced. The subsequent boycott by oil-producing nations pushed gasoline prices higher, sparked renewed interest in small cars, and brought on the worst recession since World War II. Fights broke out at gas pumps, unemployment reached 9 percent, and policy-makers in Washington searched for a way to cut dependency on imported oil. Taxing gasoline more heavily in order to reduce consumption was seen by most legislators as political suicide.

Instead, Washington wanted Detroit to find technology that made vehicles more fuel efficient. The staffs of government energy agencies devised a formula called "corporate average fuel economy" (or CAFE). Each automaker's CAFE was based on the fuel efficiency of each of its models multiplied by the number of each model it sold. Rather than dictate how much a car should weigh or how efficient specific models should be, each automaker was responsible for complying with a goal based on average miles per gallon of cars sold. Otherwise, a monetary penalty was imposed. The 1975 Energy Policy and Conservation Act required each automaker to reach an 18-mile-per-gallon CAFE by 1978 and a 20 mpg level by 1980. Later in the year, while rolling back surcharges on imported oil, Congress set 27.5 mpg as the goal the auto industry must meet by 1985.

The quickest way for automakers to boost their fuel efficiency average was to sell more small cars. Trouble was, most Americans, while favoring energy conservation, weren't buying small models — or anything else. GM's Vega, Ford's Pinto, and Chrysler's Valiant were gathering dust. On Super Bowl Sunday 1975, Joe Garagiola on behalf of Chrysler introduced the first discounts in the form of cash, meant to clear a massive inventory of unsold cars. Thus the word *rebate* entered the American popular vocabulary.

CAFE was going to cost the automakers more money than air-pollution or safety regulations. Researchers at the federal Transportation Systems Center think tank estimated that compliance would cost between $60 billion and $80 billion, more than all of the invested capital represented in the Big Three. New materials had to be found to make cars lighter, passenger compartments had to be made smaller, the more fuel-efficient front-wheel drive would have to replace rear-wheel drive on many models, necessitating reconfigured assembly lines. The burden on the industry was immense, but particularly so for Chrysler because its sales volume was the smallest.

By the end of 1975, Chrysler was once more choking on a three-month inventory of unsold cars, many covered with snow on the Michigan State Fairgrounds in Detroit. Chrysler lost $260 million in 1975 and suspended the dividend on common stock for the first time since 1933. Townsend's rescue plan for Chrysler was drastic. The warranty was extended to cover several parts that previously were excluded. He dropped several large car models and threatened to close the Jefferson Avenue plant. (Protests by the UAW, the city of Detroit, and the state of Michigan influenced him to save the plant.) For the first time the automaker dipped into its $250 million line of emergency credit. For the third time during Townsend's tenure, Chrysler dodged a bullet. In July, having had enough, he announced his retirement and nominated Riccardo to be his successor.

The economy picked up in 1976 as Chrysler prepared to intro-duce two new compact cars, Dodge Aspen and Plymouth Volare. The idea of the new models was to make fuel-efficient cars more desirable by making them more luxurious than Dart and Valiant. Initially sales of Aspen and Volare were strong. The models won jointly the Motor Trend Car of the Year Award. Chrysler returned to profitability. In 1977, the bottom fell out as the new compact car proved to be riddled with defects. A million Aspens and Volares were recalled for possible brake failure. Two more recalls were announced, one for bad hood latches, a more serious one for stalling. The Motor Trend winners received the Lemon of the Year award from the Center for Auto Safety, the consumer watchdog group founded by Ralph Nader. Aspen and Volare sales dropped off 20 percent.

To comply with new regulations, Chrysler committed itself to a $7.5 billion capital spending program to develop a series of more fuel-efficient front-wheel-drive replacements for the Aspen, Volare, and other rear-wheel-drive models. Two front-wheel-drive subcompacts, the Omni and Horizon, had already been introduced in 1977. The front-wheel-drive K car—under the supervision of Hal Sperlich, a former Ford executive who had been fired by Henry Ford II—was under development to replace Aspen and Volare in the compact class. But cutbacks by Townsend in 1970 and 1974 shrank the engineering department, leading to chronic delays. Already $1.2 billion in debt, Chrysler was paying for its capital projects out of cash flow. So when car buying stalled in 1978, losses were deep and worrisome.

As if troubles at home weren't bad enough, Chrysler's overseas empire was collapsing. The idea of building a global network had been sound. Because Ford and General Motors had done it first, however, Chrysler wound up with the leftovers: weak automakers in Britain, France, and South America. By 1976, Rootes of Great Britain had lost money in six of nine preceding years. Only a $300 million bailout from the British government prevented a shutdown. An urgent need for cash motivated Riccardo to sell unprofitable operations in Australia and Brazil during 1978. In August, Peugeot-Citroen bought Chrysler's British, French, and Spanish plants for $230 million cash and $202 million in Peugeot stock.

Chrysler's future was looking bleaker by the quarter. The Aspen and Volare were recalled again, this time for suspension problems. Production snafus kept several other models in short supply. Stockholders at the annual meeting in April 1978 approved a $200 million issue of new stock to help pay for capital projects by a narrow 58 percent majority. It didn't take a clairvoyant to see that patience with the company's mismanagement was wearing thin.

On November 2, 1978, Chrysler announced its worst quarterly loss ever, $159 million. Sales for the industry were rising but Chrysler couldn't get its act together. Previously the company had been telling

Wall Street it would be profitable for all of 1978. After the third quarter, all pretense of fulfilling that prediction was dropped.

The three-year tenure of John Riccardo, an executive who understood numbers but very little about cars or how to build them, had failed to reverse Chrysler's plummeting fortunes. But Riccardo knew he was failing and was willing to step aside. Nevertheless, he had no intention of passing the reins to Gene Cafeiro, Riccardo's number two, a man with whom he had been feuding since taking office. Chrysler needed more than mere executive competence, it needed heroism and vision.

In August 1978, just a month after Henry Ford II summarily fired Lee Iacocca, Riccardo asked him to join Chrysler. The same day that the company announced its worst quarterly loss ever, Iacocca was confirmed as Chrysler's president and chief operating officer.

THE REAL LEE

IN THE SUMMER of 1986, standing before the newly refurbished Statue of Liberty on a perfect Fourth of July as millions cheered and Ronald Reagan smiled approvingly, Lee A. Iacocca should have been a thoroughly fulfilled man.

Few American business figures, no matter how successful, ever commanded equal billing with the president of the United States. Indeed, many influential people were convinced he might make a fine president himself and were already trying to persuade him to run. He was among the few who had climbed to the top of a giant corporation, stumbled, and regained his footing. Chrysler had appeared doomed before he showed up, and now, under his leadership, was healthy. His oratory and rhetorical powers were magnificent. None before him had survived long enough at the top to tell his own tale of corporate warfare — in his own words, with his own spin — and watch it become a best-selling book worldwide.

Iacocca was among the lucky few who had overcome personal misfortune as well. A widower with two teenage daughters, he had found a new, young, adoring mate to share his later years, good health, and the personal fortune his corporate victories had brought him.

Nevertheless, he was troubled. The celebration for the Statue of Liberty, during which everyone heaped fulsome praise on him for raising nearly $300 million in private donations to restore the statue, had been marred by his squabble with the Reagan administration.

28

Donald Hodel, Secretary of the Interior, insisted Iacocca shouldn't run both the advisory committee *and* the fund-raising committee. To anyone but Lee Iacocca this might have been a minor slight. There were other irritations. His second marriage was causing tension at home. His daughter Kathi was chilly to his wife Peggy. He and Peggy argued—about money, about furnishings, about whether to live in Detroit or New York.

Chrysler seemed to be running well, but it already was clear by mid-1986 that the automaker wouldn't be able to match the previous year's record profits. And there were murmurings of unrest among Chrysler officers, engineers, and, as always, among workers in the assembly plants. The workers, having sacrificed income during the government-sponsored bailout a few years earlier, were unhappy with their pay. Designers and planners were urging him to accelerate development of new models to compete with new Japanese cars and trucks that were gaining popularity—and stealing sales from Chrysler—day by day. Others among his inner circle were urging caution: new cars cost hundreds of millions, sometimes billions of dollars, to develop and launch, and the return from that investment was uncertain. Better to take that money, they urged him, and invest it in nonautomotive ventures, where the Japanese weren't so dominant.

The Japanese: yet another worry. The media spoke of little else: Japanese quality. Japanese efficiency. Japanese service. The chant rang in his ears: Honda, Honda, Honda; Toyota, Toyota, Toyota. The sound gnawed at him, not because the Japanese got under his skin. That was true enough, and he never dared talk disparagingly about them except among his closest confederates. The press would instantly brand him a racist, and he didn't see himself as that. For the time being, the press was content to call him a "Japan basher" and "protectionist," derogatory terms used by Ronald Reagan and his bunch against anyone who wanted a government trade policy that helped American business and kept American workers off the unemployment lines. All he wanted were some sensible measures from the government to restrict importation of cars from Japan.

Couldn't anyone but him see that buying Japanese cars was turning into a national dependence that invited economic doom? Japan's so-called voluntary export restraints—negotiated under pressure from the Reagan administration in March of 1981—had been set at 1.68 million cars annually. Since then the ceiling had risen steadily, so much so that it was difficult for Chrysler to sell its own models. He never dreamed the Japanese automakers would be shrewd enough to calculate that if their exports had to be limited, they would export the more expensive luxury cars. Thank God for the 25-percent tariff on light trucks and minivans, which for the moment made those vehicles too costly to bring to the United States.

A public-policy foray against Japan had blown up in Iacocca's face. He took to the stump in the early 1980s to criticize Japanese automakers, saying they should build cars in the United States if they want to sell here. He demonized the Japanese as aliens, invaders, and despoilers of America's economic landscape. He figured that United Auto Workers–style work rules and wages would scare them away. He guessed they would never build U.S. factories in any significant number, and if they did, they'd fail badly. He never imagined that the Japanese automakers regarded manufacturing in the United States as a linchpin of their long-term strategy and that they already had plans to build assembly plants, both union and nonunion, all over the country.

As Lee gazed over the flotilla of vessels that crowded New York Harbor, including the *Queen Elizabeth II,* which had been chartered by Chrysler to fête top-selling dealers, he must have wondered if Henry Ford was at home watching the spectacle on TV. Sometimes, at moments when he was basking in acclaim, he couldn't help but speculate whether Henry wasn't sorry he had fired him. If there was any justice in the world, Henry should have made *him* chairman of Ford Motor Company instead of firing him. That's what he should be right now. In fact, this could have been Ford's party. Memories of those final, sad days at Ford pricked him now and then, injecting glimmers of anger and loss into his happy moments.

He still couldn't comprehend what he had done wrong or why Henry had rejected him. His wife actually suggested he should write Henry a thank-you note, because Henry truly had done him a favor, setting him free to win the gold ring at Chrysler. He didn't see it that way. As he explained in his autobiography, *he* was responsible for creating cars like Ford's Mustang and the Mark III, which brought honor and fortune to Ford. *He* had performed like a devoted son, doing everything in his power for Ford and, ultimately, for Henry's family. But nothing was enough. Imperious Henry would never accept as chairman of his company an Italian, a small-town striver, a guy who didn't give a damn about Grosse Pointe debutante balls or the Social Register.

Lee felt that Henry had acted irrationally, though he sometimes wondered if his old boss might have detected a flaw in him that no one else saw. The flaw was insecurity. But in Iacocca's eyes it was as much a strength as a weakness. That glimmer of self-doubt, revealed to very few people in his life, was for Lee the grain of sand in the oyster, the force that drove him again and again to prove that he belonged.

By the time Chrysler's chairman reached his pinnacle, that shining moment in New York Harbor, the basic outline of his life had become as well known to many Americans as George Washington's.

He was born in Allentown, Pennsylvania, on October 15, 1924, the son of Italian immigrant parents. His father was a hot dog vendor, his mother a housewife. A talented pupil and debater, he missed military service in World War II due to a medical deferment, caused by a childhood bout with rheumatic fever. Early on he learned about feelings of discrimination and inadequacy. By his own account, Iacocca was the target of anti-Italian slurs from teachers and schoolmates. In addition, having experienced "disgrace" as a result of the draft deferment, he resolved to restore his self-esteem by excelling in his studies.

His father grew prosperous enough to send him to college. Iacocca breezed through his B.A. in mechanical engineering at Lehigh University in less than three years, earning high honors and working as a layout editor on the school newspaper. While in college, driving a 1938 Ford his father had bought for him, he decided he wanted to work for the company that had built that car.

In 1944 he was one of fifty U.S. college graduates offered a job by Ford. It was an appointment coveted by many, a chance to work for one of the nation's premier companies. Iacocca, however, decided to take another two years—long enough to complete a master's degree in mechanical engineering at Princeton—before joining Ford, in 1946. Naturally, once he was at Ford, the company hoped to transform Iacocca into an automotive engineer. But after spending nine stultifying months designing a clutch spring, the young trainee realized, astutely, that sales and marketing, not engineering, was where the action was. In the postwar boom, quality and performance weren't paramount issues. Customers were lined up to buy every car Detroit could build; the trick was snagging the customers before General Motors or Chrysler did.

Ford wasn't willing to transfer Iacocca to sales; his bosses, however, let him know that if he wanted a sales job somewhere in the company, he would have to sell himself to a supervisor of a zone or a region. The necessity of "applying" for a job inside a company that already employed you was a telling demonstration of the feudal nature of large corporations. Qualifications mattered less than whether you were acceptable to the boss. The corporation overnight had become the dominant expression of American business. You learned its methods or you moved on. Departments were small realms run by a boss. To a worker, no one was more important than the boss; loyalty to the boss ranked alongside, and sometimes above, loyalty to the company. As an engineer, Iacocca would be something of a square peg in a sales department, so he had some convincing to do if he was to get a boss to take a chance on him. He persuaded the supervisor of Ford's Chester, Pennsylvania, district office and soon demonstrated a natural talent for pitching and shmoozing Ford dealers in eastern Pennsylvania.

By 1956, after a decade on the job, Iacocca managed to attract the attention of Ford brass at the headquarters in Dearborn, Michigan. He had a gift for stunts and slogans, and a powerful speaking style. Iacocca knew how to reduce ideas to their essence. Instead of showing a training film on automotive safety to fellow salesmen, he dropped an egg from a stepladder to demonstrate how padded dashboards could prevent injury. He conjured a sales campaign for eastern Pennsylvania called "56 in 56"—based on a down payment of $56 for a new car—that was adopted for nationwide use. So when his mentor, Charlie Beacham, was promoted from eastern regional headquarters to national headquarters, he brought Iacocca along as one of his subordinates.

Here was another lesson that wasn't lost on Iacocca: a boss never sallies forth without a retinue of vassals to support and defend him. Before venturing to Dearborn, Beacham needed a loyal staff for his new job. Iacocca, whose own career was taking off, was learning more about tribal politics: make your boss look good and he'll take care of you; attach yourself to a rising star and never venture into foreign territory without a solid phalanx of loyalists.

Fortunately for Iacocca, Beacham grew quite chummy with one very important resident of Ford headquarters, Robert S. McNamara, vice president of the division selling Ford cars and trucks. McNamara, a brainy former air force officer who later would become Secretary of Defense under John F. Kennedy, had been a member of the "whiz kids" hired by Henry Ford II to revive the failing automaker in 1946. Himself cold and aloof, McNamara nevertheless appreciated Iacocca's verbal flair. Iacocca had been an adherent of the Dale Carnegie method, never afraid to extemporize, even if he had little to say. Selling was largely motivational, a lot of flash and dash. McNamara wanted Iacocca to become more analytical. He ordered Iacocca to write down his ideas as a way to test whether they were as good as they sounded.

Lee was a quick study. He learned how to run a meeting, how to express his views, and how to negotiate among the competing factions inside Ford. Of these factions, the most powerful group was the financial analysts and accountants—the bean-counters. This was McNamara's group, which included many M.B.A.s from Harvard and other prestigious business schools. Their passion was numbers, ratios, and return on investment—not cars.

Another powerful group was the product planners, also known as the "car guys." Most of them were engineers, and their mission was to conceptualize the cars of the future, build the prototypes, test them, determine if they could be manufactured profitably, and then sell their ideas to Henry Ford. Every new model was a major gamble that spanned several years, cost hundreds of millions of dollars, and consumed tens of thousands of hours of engineering analysis. Every new car carried with it the possible huge penalty of lost opportunity: while one automaker was building an unsuccessful car, rivals were stealing customers with successful ones.

Car guys and bean-counters were natural antagonists. Car guys argued that expenditures of huge sums were vital to keep abreast of technology, fashion, and the driving public's growing appetite for creature comforts. Bean-counters were paid to be skeptical. Every expenditure had to be supported by research, every investment had to be justified by the promise of a profitable return.

The factory rats were the lowliest group of all. The details of manufacturing didn't interest most auto executives, except insofar as they affected bottom-line profit or the ability to ship cars fast enough to dealers. From a career standpoint, a managerial job in the factory tended to be a ticket to nowhere. In the view of most executives, factories were hot, dirty, dangerous places to be avoided; they existed solely to perform the grim tasks associated with bending, welding, and painting metal. The workers belonged to their own clan, the United Auto Workers, and there was little fraternization or respect between union members and "suits."

Iacocca belonged to yet another group, the "sales guys." In his earlier days in the Chester, Pennsylvania, office, his job was to meet with dealers, listen to their complaints, soothe their anxieties. As he moved up the ladder, he supervised the daily *Sturm und Drang* of sales from a corporate perspective: the marketing plans, reports of which models were selling, which weren't (and why), the business of distributing vehicles equitably after they left the plant, as well as the advertising, pricing, research, and so forth. Sales guys rarely cracked the top executive echelon, though they were crucially important to an automaker's success.

During the three decades following World War II, the antagonism among different constituencies within the automakers, and even the subdepartments within groups, was accepted as a matter of course. Though executives routinely paid lip service to "teamwork," they encouraged intramural competition insofar as it identified and tempered strong leaders and forced aside the chaff of second-rate ideas. True, everyone first and foremost worked for Ford and derived identity from the blue oval. But within that context, departments were like warring tribes.

Ford tried to hire only bright, ambitious, and highly motivated managers, and from their first day on the job, they were expected to compete fiercely among themselves for promotions, raises, and kind words from the boss. Anyone who didn't dream of one day making it to top management certainly never would, and probably didn't belong. Selling an idea or design took brains and articulateness; it also required the aggressiveness to challenge the author of any competing idea. To control and channel this aggression called for executives who could make sure the sparring remained constructive. (Don Petersen, who became chairman of Ford in the 1980s, wrote after his retirement that he considered quitting several times early in his career because of the intimidation and antagonism he observed among co-workers.)

In 1960, Henry Ford chose McNamara to be president of the Ford Motor Company. McNamara, in turn, chose his thirty-six-year-old wunderkind, Lee Iacocca, to succeed him as head of the Ford division. On the day he took over, there were people at Ford who had barely heard of him and many who had never seen him. By Iacocca's own estimate, he had vaulted over at least a hundred older and more experienced people to reach the exalted post of corporate vice president, a member of the Ford elite. Under his command were 11,000 engineers, planners, managers, and secretaries, and more than a few of them believed that Iacocca's promotion was not deserved. To be accepted as leader Iacocca had to win respect and allegiance, a goal he accomplished with the development of the legendary Mustang.

Detroit's oral and written history is replete with differing versions of how the Mustang was created and who most deserves credit for its creation. There's little doubt, however, that Mustang has been seen as Lee Iacocca's achievement. It was Iacocca who assembled the planners and engineers, a team led by Don Frey and Hal Sperlich, the so-called Fairlane Committee, to develop the basic layout. It was Iacocca who asked Gene Bordinat, head of styling, to produce a special exterior design, which Bordinat accomplished by conducting a contest among three of his best designers. Above all, it was Iacocca who succeeded in persuading Henry Ford to proceed with the project in the face of doubts from his beloved bean-counters.

The Mustang mirrored the youth and vitality of the early 1960s. Mechanically, it was little more than a Ford Falcon, the utilitarian car that had been McNamara's pet project, with a restyled interior and exterior. Unlike the Falcon, which embodied McNamara's innate frugality and practicality, the Mustang conveyed quickness, energy, and style. One of the most popular cars ever, the Mustang was a financial home run for Ford, registering more than $1.1 billion in sales for the automaker between 1964 and 1966. The Mustang's phenomenal success also landed Iacocca (and the car) on the covers of *Time* and *Newsweek* during the same week in April 1964. From that week on, all the petty quibbles about which *individual* was responsible for the Mustang were over: Iacocca's paternity was established beyond question, at least in the public consciousness, and with a certainty never before seen in Detroit.

If Henry Ford had any misgiving about Iacocca's celebrity becoming too large and unseemly for his family's company, about it possibly overshadowing the name on the outside of the building, he didn't express it publicly. On the contrary, he appointed Iacocca head of cars and trucks for Ford and Lincoln-Mercury in January of 1965 and gave him a seat on the board of directors. At age forty, Iacocca was one level below president, the automaker's top operating job and one of the single most powerful spots in American business. That he burned for the job, ached for it, said nothing derogatory about him. Raw ambition was a qualification for auto executive; ambition supplied the sort of dynamism necessary to create cars like Mustang.

McNamara left Ford to serve in the Kennedy administration, and was succeeded by John Dykstra and then another whiz kid, Arjay Miller. In 1968 Henry Ford abruptly removed Arjay Miller as president and made him vice chairman, an advisory post. Henry couldn't seem to find a president who satisfied him. Walter Hayes, a longtime associate of Ford, opined that Miller may have been "too fine an intellectual" to deal with the growing political and social demands for safer, cleaner cars. Ralph Nader had abruptly altered the environment for automakers, forcing them to think about tougher regulations and what that meant. Prior to removing Miller, Henry Ford recruited Semon E. Knudsen, a senior GM executive with the nickname "Bunkie," as a replacement. The choice made sense, given Henry Ford's admiration of GM's management proficiency and GM's overwhelming dominance of the U.S. automotive market.

Miller never betrayed any bitterness or disappointment about his removal from the presidency. He understood that Henry Ford hired and fired as he pleased, with little pretense of sober deliberation. Although Ford had been a publicly held company since the late 1950s and technically its board of directors were supposed to supervise such matters, Henry Ford II's influence was obvious and overwhelming. When you came to work at Ford you had to understand that Henry, for better or worse, was the absolute monarch. Whether or not he did the right thing, the issue for a senior executive was to ensure that the company was well served and to understand that there was no court of appeal for Henry Ford's judgments.

Iacocca believed that he was a far better choice for president than Knudsen in 1968. But to discuss the subject with Henry Ford was unthinkable. Iacocca had no intention of swearing allegiance to Bunkie, nor of biding his time. In his own memoirs, Iacocca describes Knudsen's problem at Ford as "an inability to keep up," the final straw being Knudsen's habit of barging into Henry Ford's office without knocking. Iacocca's assessment of Knudsen was hardly disinterested. According to numerous recollections by Ford executives of that time, Iacocca undermined Knudsen from the moment he arrived, routinely countermanding his orders and disparaging his abilities to Henry Ford.

Knudsen arrived at Ford without allies. It was a tactical mistake on his part and Henry Ford's. In less than two years he was rendered ineffective by Iacocca's powerful tribe of loyalists.

Henry, according to Walter Hayes, recognized that there was potent resistance to his appointment of Knudsen. Iacocca had built an impressive coterie of admirers, including William Clay Ford, Henry's brother, and several directors. And Ford's chairman couldn't completely fault Iacocca for acting in the manner aggressive executives were *expected* to act; but he wasn't prepared to crown Iacocca's tactics by immediately handing him the presidency. Instead, he established an executive triumvirate composed of Iacocca, as president of North American operations, and two others, the presidents of international and nonautomotive operations. The triumvirate lasted barely a year, Henry Ford finally conceding the structure made little sense. On December 10, 1970, Iacocca's ambition, talent, and guerrilla tactics paid off when he was named president.

From 1971 to 1975 the Ford Motor Company prospered, though a sea change was under way that would alter the nature of the automobile business. The Clean Air Act of 1970 was followed by the Yom Kippur War in 1973 and a jump in energy prices. Henry Ford, who had shown some reservations about Iacocca's business judgment, decided in 1975 to replace him, as he had replaced Miller, Dykstra, and others. He respected Iacocca's uncanny sense for identifying consumer tastes, choosing designs that appealed, and marshaling the large number of talents it took to create vehicles. He was less sure that Iacocca possessed the managerial breadth to integrate the complexities and demands of an enormous corporation with American society's spiraling concerns about the environment, safety, and a blizzard of other issues.

During his term as president, Iacocca probably hobbled his ambition to succeed Henry Ford by reveling a bit too excessively in the splendor that came with his president's title. He traveled in regal style in a lavishly decorated Boeing 727. He lodged in the Ford suites at the Waldorf in New York and at Claridge's in London. He regularly caroused in New York with celebrated buddies like singers Frank Sinatra and Vic Damone, George Steinbrenner, and limousine operator Bill Fugazy. Henry Ford didn't like Iacocca's taste for high-profile glitziness. Nor did he like Alejandro de Tomaso, the Italian designer and Iacocca pal who created the Pantera for Ford, a sexy sports coupe that didn't sell well, lost millions, and raised concerns among safety regulators. Iacocca's friendships made Henry nervous. Being secretive and mistrustful by nature, he suspected some of Iacocca's friends, particularly Fugazy and Tomaso, were exploiting their association with Iacocca to get rich from Ford business.

Iacocca failed to comprehend that he was never Henry's protégé nor an *ex officio* family member, but a hired hand who would be replaced at the boss's whim. The discreet tack would have been to play down the perks that went with his rank, to act humble. Henry preferred a president who displayed some insecurities, a worrier who would call him before making an important decision, who might suggest *selling* the company jet because it hurt the company's profit, the family's profit. Henry was offended by a president, a hiree, who acted as though he was entitled someday to inherit the whole enterprise. Lee never understood; and it's not surprising if he felt puzzled, betrayed, angry and, finally, vindictive when Henry Ford began sending signals that he was on his way out.

Henry Ford had a case for replacing Iacocca on business fundamentals alone. Despite the success of the Mustang, Ford's market share in the United States was falling in the mid-1970s and overall profits were weak. The Pinto had been a disaster. Henry's age and declining health made him view a successor to Iacocca as a pressing matter. Having split with his second wife at the age of fifty-eight, and having already suffered his first bout of heart trouble, he began considering a broader reform of Ford's management, preparing for the day when he was no longer there.

By the fall of 1975, Henry Ford had decided definitely that he wanted to fire Iacocca and replace him with Bill Bourke, chairman of Ford of Europe. Ford's outside directors and others counseled Ford to go slowly and to avoid a revolutionary upheaval in the ranks. They remembered the Knudsen affair and didn't want a repeat. Iacocca's men were everywhere. Besides, the directors weren't entirely convinced that Henry Ford had a legitimate reason for firing Iacocca. The turmoil would not be healthy for operations. So Henry instead took the measured step of promoting Bourke to run North American operations.

In the meantime, Henry Ford was hospitalized for the first time with angina. Spurred by intimations of mortality, he decided a few months later to hire McKinsey and Company to help him figure out how to solve the riddle of top management succession. As Iacocca would write in his autobiography, he saw the hiring of McKinsey "an act of war" against him. Iacocca had many supporters, among them members of the board of directors who thought it was a terrible idea to fire Iacocca at a time when Henry Ford was having heart trouble. Yet Henry remained determined to devise a succession plan that didn't leave Iacocca in charge.

In April of 1977, Henry Ford announced, according to the recommendation of McKinsey, the creation of an office of the chief executive. He was to remain chief executive through 1979 and chairman through 1982; Phil Caldwell was vice chairman, and Iacocca was president. For the moment, Bill Bourke stayed in place. Reporters trying to read the tea leaves pressed Henry Ford to concede that, in his absence, Phil Caldwell, not Iacocca, was in charge.

In all this shuffling, Iacocca retained a very important job, but in the larger scheme of things, he had been eclipsed by Caldwell. Someone else might have received the developments as immutable, a sign that the game was up. Iacocca, in his combative style, wasn't about to surrender. He hadn't knuckled under to Knudsen and, though he didn't like working under Caldwell, he realized his position now was too precarious for open rebellion. But Iacocca wanted to show that trying to fire him was proof of his boss's incompetence. Henry Ford, in the meantime, had himself learned something about warfare from the Knudsen episode. He fired one of Iacocca's key allies, Hal Sperlich. Another, Paul Bergmoser, was forced into early retirement. Ford's relationship with Fugazy, who had organized lavish trips for dealers, was terminated. Iacocca's men were on notice: Henry was out for blood and wouldn't tolerate any dissent to his rule.

By 1978 Iacocca's tactics had grown desperate. Two class-action lawsuits initiated by the notorious litigator Roy Cohn against Henry Ford for alleged misapplication of corporate funds and bribery of foreign officials were a great personal embarrassment to Henry. The legal actions were flimsy and ultimately collapsed. But Cohn was a great chum of Fugazy's, and while there was no evidence that Iacocca had anything to do with the lawsuits, it certainly looked suspicious to Henry Ford.

The directors and Bill Ford, Henry's brother, still regarded Iacocca as an important talent and tried to affect a reconciliation. Henry was adamant; he still wanted Iacocca out. While Ford was away on a trip to the Far East in June of 1978 (taken in part to meet with Chinese leaders on the subject of growing Japanese influence), Iacocca flew to Boston to lobby directors about keeping his job. In Iacocca's view, Henry was losing it and, therefore, no longer able to understand how important Iacocca was to the company. For some time his comments about Ford's chairman had been growing more caustic and indiscreet: he told people that the old man was senile, that he drank too much and had no business running a public company.

Whether or not directors sympathized with Iacocca was irrelevant. They served at the pleasure of the Ford family and would instinctively support Henry in the crunch. Arjay Miller, according to Hayes, informed Henry about Iacocca's meetings during his absence, assuring his boss that he now had incontrovertible evidence of Iacocca's treachery and was justified in firing him. Later on, in his autobiography, Iacocca would insist that directors had invited him to Boston, that there was nothing secret or disloyal about their meetings.

Nevertheless, on July 12, 1978, Henry Ford attended the organizational subcommittee of the board, prepared to resign if directors didn't endorse Iacocca's firing. According to Hayes, the matter was decided without discussion. At 3:00 P.M. the next day, Henry Ford summoned Lee Iacocca to his office and, with brother Bill at his side, told him he was through.

The greatest auto executive of the decade had met his match in Henry Ford, just as every athlete, politician, tycoon, and movie star one day discovers he has been bested by someone quicker, smarter, richer or better-looking. Henry had been the stronger, the wilier; and, by virtue of his last name, he commanded the heavier artillery. The terrible events of that day, suffered at the hands of his boss, were received by Iacocca on a purely personal level. The magnitude of bitterness these events engendered in him, however, was unjustified. For thirty-two years he had waged corporate warfare well, even brilliantly. He knew precisely how brutal life in an auto company could be, for he had dispatched not a few men in a manner as cold and cutting as he had been dispatched. This was the Darwinian culture of Ford, and of most large U.S. corporations: the battle of ideas, the assertion of personal power, the struggle for scarce resources, the constant cycle of domination and submission. Iacocca had operated within this system willingly. He understood, as Hayes astutely put it, that "when business becomes a bull ring it is hard to concentrate on matters of life rather than death."

With the helicopters, the balloons, the tall ships swirling about him, as much the center of attention that Fourth of July as Lady Liberty herself, these memories must have been very much on Lee Iacocca's mind. In the presence of such a spectacle, it was hard not to imagine that he could be elected president of the United States if he wished it. He believed so and told his friends. Anyone who thought Iacocca would accept the insulting treatment that Henry Ford had laid on him and retire meekly to Palm Springs had underestimated his rage and his desire for vindication, and therefore could not begin to imagine the will to greatness those emotions could bring out in him.

BAILING OUT

FROM A MOMENT in the summer of 1978, Lee Iacocca was America's most celebrated unemployed man. Detroit cognoscenti were betting how soon it would be before he surfaced at Chrysler. Chrysler was in desperately bad shape, and General Motors didn't hire outsiders. Add to the broth Iacocca's desire for vengeance and the fact that the only thing he knew was the auto industry, and the match made perfect sense.

By late summer, negotiations between Iacocca and Chrysler directors had begun. From the outset he let them know that the only way he would come was as chairman and chief executive. To his surprise, no power plays or back-channel pressure was necessary: John Riccardo had had enough and was willing to step aside within a year. Riccardo was tired; and he knew Chrysler's financial situation was far more grim than Iacocca or the outside world realized. Indeed, David Halberstam in *The Reckoning* said he believed that Iacocca never would have joined Chrysler if he'd known how bad things were, a point Iacocca conceded in his own book.

From Iacocca's perspective, Riccardo appeared to be in over his head and had to turn the automaker over to someone who could make it perform. (Later on, Iacocca would gloat that he heard Henry Ford was drinking heavily after the announcement of his welcome at Chrysler. If the story was true, the drinks probably were celebratory since Ford executives, including Iacocca, had long considered

Chrysler a marginal enterprise whose eventual demise was a sure thing. Ford executives fretted about GM but never about Chrysler.)

Initially, Iacocca was confident that his innate sense for matching consumer tastes with desirable cars and his leadership skills in rallying managers and engineers would be enough to turn Chrysler around in short order. Chrysler still enjoyed a reputation for engineering excellence, despite having been tarnished somewhat by numerous recalls in the 1970s for poor manufacturing quality. He didn't have a deep or thorough knowledge of banking or finance, so the condition of the balance sheet didn't worry him. He presumed the bean-counters could explain in due course, as they always had, what cutbacks the automaker needed to make money again.

As for Chrysler's car designers, Hal Sperlich, Iacocca's old ally who had joined Chrysler a year earlier after being fired by Henry Ford in 1976, was working doggedly on plans to build the K platform. The K platform was the mechanical basis for a series of small, front-wheel-drive cars (carmakers assigned arbitrary alphabetical designations to new families of cars). As an early proponent of front-wheel-drive technology—and of smaller cars in general—Sperlich had clashed frequently with Henry Ford. Henry had wanted to simply take bigger cars and shrink them down, thus saving money. Sperlich, who was something of a purist, argued for entirely new (thus expensive) designs that would improve fuel efficiency.

On November 2, 1978, Iacocca joined Chrysler as president, with the understanding from Riccardo and the board that he was to become chief executive by 1980. Adept at realpolitik, Iacocca's first step was to swiftly purge most of Chrysler's top executives and to install his own loyalists. He wasn't about to fall into the same trap Bunkie Knudsen had, blundering into hostile territory without a bodyguard. Moreover, he intended to filch some of Ford's very best executives, deriving particular joy from the thought of Henry Ford blowing a gasket.

When it came to persuasion, Iacocca had few peers. His first catch was Gerald Greenwald, the president of Ford of Venezuela. Greenwald was a former finance specialist who had once worked for Iacocca at Ford and had wished to broaden his management portfolio beyond finance. Jerry and his wife, Glenda, were skeptical about moving to Chrysler. With his understanding of figures, he knew better than Iacocca how desperate things were and realized that he would be once more cast in the role of finance man. Glenda liked their life in Venezuela, away from the dreariness and maneuvering of Detroit. But it also was a chance to be a bigger fish, and her husband believed Chrysler was ultimately salvagable. Moreover, Iacocca tantalized him with the prospect of being promoted to president in relatively short order.

Here was the beginning of Iacocca's "Gang of Ford." Greenwald recruited Steve Miller, another finance man, from Ford. Iacocca lured Gar Laux, a former general manager of Ford's Lincoln-Mercury division who had departed to venture into automotive retailing with the golfer Arnold Palmer. Laux's new job was as head of Chrysler sales and marketing. Paul Bergmoser, who had been a vice president of purchasing and an Iacocca crony, was named president. Ben Bidwell had left Ford to run the Hertz rental car company; he joined Chrysler in 1983. Dozens of other officers and managers, intrigued by the chance to join Iacocca — and offered better jobs and salaries — were recruited away from Ford.

A serious malady at Chrysler, apart from financial weakness and haphazard management systems, was old-fashioned, second-rate manufacturing. The factories were ancient and in bad repair. Chrysler's tools and machines were outdated and inefficient. And its painting technology was out of date, producing finishes that didn't look nearly as attractive, or wear nearly as well as, those of the Japanese or of the domestic competition. Workers complained of water leaking from the roofs and of oil mists that made floors slippery and fouled the air. (Exposure to hazardous fumes was a particularly bad situation in many U.S. auto plants, including those of Chrysler. Until the late 1980s, for example, many Chrysler workers were exposed to high concentrations of lead and arsenic in the air from the soldering guns used to close gaps in metal. Workers had complained repeatedly, to no avail. In 1987 the federal government imposed a $1.5 million fine on Chrysler for exposing workers to lead and arsenic at its Newark, Delaware plant — the largest penalty at that time ever levied against an American corporation for health and safety violations.)

Manufacturing executives begged for more investment, but for years they had been starved. Line workers were frustrated and dis-spirited. They were constantly showered with pious lectures on the importance of building quality cars but simultaneously harassed to move cars as fast as possible off the assembly line, regardless of the number of defects, so they could be shipped to dealers. As Chrysler's financial condition had deteriorated, the emphasis on production over quality grew more intense.

Iacocca chose Steve Sharf to run Chrysler's factories. Sharf, who had been general manager of Chrysler's engine operations, possessed an unusual background for an auto executive: he had learned diemak-ing in Germany before World War II. A Jew, he had managed to hide from the Nazis in Berlin during the war and emigrated to the United States in 1947. Starting as a skilled tradesman at a Ford factory in Buffalo, he worked his way up through the ranks, first at Ford and then at Chrysler. Having spent most of his life among factory workers and machines, he knew manufacturing intimately. He spoke English with a heavy German accent and possessed none of the polish of the upwardly mobile executive, but Sharf was extremely effective in promoting improved manufacturing methods. He was a man who cared little for corporate politics and was willing to speak up when more money had to be spent for machines or better working condi-tions.

In 1980, Iacocca recruited Dick Dauch, a former GM plant man-ager who had moved to Volkswagen. Dauch was initially responsible for Chrysler's components plants, Mexican operations, and a small number of nonautomotive interests, such as marine engines. Shortly thereafter, Dauch was named vice president in charge of manufactur-ing, Sharf's top lieutenant.

Standing six feet three inches and weighing 235 pounds, Dauch was an intimidating presence who loved to regale subordinates with grandiose tales of his exploits as a football player at Purdue and re-mind them of the relative youth at which he was promoted to plant manager at General Motors. During meetings he squeezed an exercise spring constantly, in order to keep his grip powerful for crushing handshakes. In job interviews Dauch gave much weight to participation in sports, particularly the more violent ones like hockey, football, and boxing. They were an odd pair, Sharf, the immigrant Jew who had survived the war underground, and Dauch, the farmer's son from Ohio who endlessly rhapsodized on his love for knocking heads.

Flanked by a retinue of assistants, Dauch roamed the aisles of Chrysler factories to keep an eye on operations. He didn't ride a cart like most executives because he didn't want workers to think he was soft. Fear and respect, Dauch believed, were among the most effective tools for motivating subordinates. Dauch knew times were changing, and management experts were preaching sensitivity and listening skills. But he also believed instinctively that unless foremen and manufacturing supervisors were afraid of him, they might feel free to deceive him about what was really happening in the factories. He staged numerous inspection tours, varying his route so his arrival at a work station was a surprise. Workers reacted by devising signals so they knew when "Coach Dauch" was headed their way. "A lot of people tried me out," he later said of his experiences at Chrysler, "but they learned that plants did close, people did lose their careers, and we in senior management were going to hold people accountable."

Sharf considered Dauch a fairly ridiculous figure, but Iacocca was particularly impressed by the man he referred to as "that German." For Dauch proved himself able — by sheer force of will, it seemed — to meet production quotas, no matter what or who stood in his way.

While Iacocca was building a network of loyalists in Detroit, Riccardo was buying time in Washington. In December of 1978, he met with President Jimmy Carter's top aide, Stuart Eizenstat. Chrysler was short about $2 billion it needed to undertake capital projects to meet tougher air-pollution, safety, and fuel-effiency rules coming into effect. Efforts at securing federal aid weren't going well: Chrysler had already been denied a loan of $250 million from the Farmers Home Administration to help build a transaxle plant in Indiana. Riccardo told Eizenstat that Chrysler wasn't looking for a bailout, only some temporary help in meeting federal rules, and suggested a two-year relaxation of new rules that required heavy investment on the part of Chrysler. That way, he explained, the automaker was sure to have enough capital to bring a new line of fuel-efficient compacts, the K cars, to market in 1981. Eizenstat listened.

In early 1979, Chrysler announced a $205 million loss for the previous year. Agencies that evaluated the safety of debt securities immediately downgraded Chrysler bonds and preferred stock — a signal that the automaker was going to have trouble selling securities to finance operations and car sales. To raise cash and improve its balance sheet, Chrysler decided to sell assets. First to go was the remains of its European subsidiary, sold to Peugeot for $80 million. Other foreign operations were sold for smaller amounts. Townsend's vision of a foreign empire faded into memory.

On May 29, 1979, Riccardo announced the shutdown of Dodge's main manufacturing complex in Hamtramck, adjacent to Detroit's downtown. Dodge Main, which employed about 5,000 workers, was an important underpinning of the region's economy. The seventy-year-old factory had been the principal reason for Walter Chrysler's interest in buying Dodge in the 1920s. Over the years it became more and more inefficient as manufacturing technology advanced, but it turned into a truly unacceptable expense in the face of Chrysler's anemic sales and excess plant capacity.

For low-skilled Detroiters, many of them African-Americans, a job at Dodge Main was a ticket to middle-class prosperity. Many had spent decades in the plant, applying their paychecks to their homes, boats, vacations, and children's education. Periodic layoffs during recessions weren't a problem. Union and government unemployment benefits provided a partial safety net. Low-paying jobs filled the gap. The permanent closing of Dodge Main's gates was something scarier; it spelled a return to poverty, the disintegration of families, communities deprived of their focal point.

The financial losses and the lack of government response to entreaties for help made Chrysler look very much like a basket case. At Chrysler's board meeting in Syracuse in May 1979, Felix Rohatyn, the renowned investment banker from the firm of Lazard Frères, suggested that the directors put him in charge of a special committee with power over management. Riccardo had hired Rohatyn to explore restructuring strategies. Rohatyn envisioned reorganizing Chrysler the way the Municipal Assistance Corporation had reorganized New York City in the mid-1970s. Riccardo was now furious with Rohatyn, whom he perceived as engineering a power play to aggrandize himself. The board backed Riccardo and dismissed Lazard Frères, replacing the investment bank with Salomon Brothers. (A few months later, according to Robert Reich and John Donahue in *New Deals,* a book about the Chrysler bailout, Rohatyn told Eizenstat that Chrysler was doomed and that the Carter administration would be wise to leave it to the bankruptcy court.)

The day after Rohatyn's appearance in Syracuse, Lee Iacocca made his first address to the shareholders. Upbeat about Chrysler's future, Iacocca blamed external factors — energy prices, high interest rates, government regulation, and imported vehicles — for Chrysler's troubles. There were no direct references to management mistakes or poor quality vehicles. However, Iacocca vowed to abolish the sales bank, the previous management's mechanism for storing excess, unordered cars and then pushing them on the dealers.

The very next month, aware that the second quarter had been dreadful, Riccardo journeyed once more to Washington. This time Riccardo visited the White House, and he went in the company of Senators Donald Riegle of Michigan and Thomas Eagleton of Missouri, many of whose constituents worked in Chrysler factories in St. Louis. Riccardo painted a grim picture for Eizenstat and W. Michael Blumenthal, then the secretary of the Treasury, suggesting several possible government remedies to reduce financial pressure. Riccardo again asked for a delay of air-quality and other regulations that would allow Chrysler to delay investments. Once more, Eizenstat promised nothing. In a separate meeting, Douglas Fraser, president of the United Auto Workers Union, suggested to Eizenstat that the government buy a stake in Chrysler. Fraser's suggestion, smacking of British-style socialism, was shot down immediately.

Chrysler meanwhile tapped its $560 million line of backup credit in the summer of 1979, a move that was viewed by bankers with alarm. The backup credit was meant only as a short-term safety net for transactions of commercial paper, financial obligations sold by Chrysler to raise money from public investors. The bankers routinely extended the credit, never dreaming it might be used for operations or capital spending.

At the end of July, Chrysler announced a $209 million second-quarter loss, forcing Chrysler executives to sharpen their pleas and the Carter administration to consider more seriously the question of whether to help Chrysler or let the bankruptcy laws take their course. The laws governing business failure had changed a great deal since the times when a man's family could be sold into slavery to fulfill economic obligations. U.S. bankruptcy law, which evolved in the nineteenth century, allowed failing companies to continue operating under supervision of a court-appointed trustee. The law adhered to the principle that companies might be saved if they are reorganized, thereby maximizing the value of their assets; still, in the event of failure, creditors must be paid in a just and orderly fashion.

In the twentieth century, with the rise of large corporations, the stakes in a financial crisis have grown proportionately. In a large-scale corporate bankruptcy, hundreds of thousands of jobs are at risk, as are entire communities, charities that depend on corporate donations, and overall economic confidence. "Too big to fail" was a phrase invented to describe the company or bank whose sudden demise might topple an entire industry or trigger a massive loss of consumer confidence. To the extent that government controlled certain economic levers, it also possessed the means to resuscitate an ailing company, at least in the short term. The decision to use those means, to alter the course of free enterprise, fell to politicians whose own futures depended on their wise use of the controls at hand.

Three high-profile government bailouts during the 1970s shaped the Carter administration's policy for dealing with Chrysler's ills. In the 1970 bankruptcy of the Penn Central Railroad, politicians worried that commuters faced the possibility of no transportation; in the 1970 federal rescue of Lockheed, the Pentagon had to consider who would build the C-5A aircraft and the Trident missile. New York City skidded to the brink of financial collapse in early 1975 when revenue was insufficient for operations and the city had lost access to credit as a result of chronic profligate spending and borrowing.

In the case of Penn Central, the federal government carved a private company from the old and troubled railroads. The government, at a cost to taxpayers of about $5 billion, shouldered the cost of carrying freight to small towns, subsidizing shipping rates and high union wages. With Lockheed a significant precedent was created in 1971 when Congress approved a $250 million federal loan guarantee, accompanied by government oversight of Lockheed's management. After Lockheed vindicated supporters of the bailout by repaying the loans, the stage was set for a $1.65 billion federal loan guarantee for New York City in 1978.

In the matter of Chrysler, political sentiment was adrift in the summer of 1979. The White House wanted to help, but Chrysler initially wasn't interested in a loan guarantee. Riccardo wanted a $1 billion tax refund, in cash, in advance, against previous losses. He argued that Chrysler, the smallest of the Big Three, had been in part a victim of excessive government regulation. G. William Miller, the new secretary of the Treasury who had replaced Blumenthal, unequivocally opposed the idea. In August of 1979, with finances worsening by the day, Riccardo met with Eizenstat and other White House officials, ready to discuss a loan guarantee instead of an advance tax refund.

Prior to championing a loan guarantee, the Carter administration had wanted Chrysler to submit a detailed operating plan showing how the automaker intended to recover. Chrysler's management had been so chaotic that it not only lacked detailed plans but was keeping only sporadic statistics. Within three weeks, Greenwald and Treasury officials drew up a six-point blueprint for recovery, including new management with Iacocca in charge. Chrysler would reduce the number of car-body styles from five to three, conserving cash by extensive "reskinning"—developing new exteriors on existing mechanical platforms. This, indeed, was Iacocca's forte, creating new models with minimal investment: a set of fake wire wheels here, a vinyl roof there, a new name supported by a clever advertising campaign—voilà, a "new" model. Central to the operating plan was the K car for 1981, Hal Sperlich's front-wheel-drive design. The K car needed substantial additional investment, which would not be forthcoming without a sharp reversal in the automaker's fortunes.

By the fall of 1979, the White House was sympathetic to loan guarantees, but action was impossible without a political consensus in Congress. So Wendell Larson, Chrysler's public relations chief, hired a small army of Republican and Democratic lobbyists, recruited Michigan's legislative delegation, and drafted nearly 2,000 Chrysler dealers and partisans of organized labor to sell the Chrysler bailout to skeptical congressmen and senators. The lobbyists argued that a conventional bankruptcy would destroy Chrysler's already weak vehicle sales, jeopardizing its chances for recovery. The cost to government of unemployment compensation and lost tax revenue from a bankruptcy would be enormous. Moreover, they argued, previous bailouts had worked.

William Proxmire, chairman of the Senate Banking Committee, summarized the ideological opposition to helping Chrysler: "If we provide loan guarantees to Chrysler, we will be saying, in effect, to every business in the country that it doesn't matter if you no longer make products that enough people want to buy, it doesn't even matter if the federal government has no direct stake in your continued existence. None of this matters so long as you are big enough and can muster enough interest groups to fight your cause in Washington." Proxmire, who had opposed the bailout of New York City as well, favored a bankruptcy filing.

Proxmire had a point. Numerous interest groups, individuals, and eminent economists opposed aid to Chrysler on the grounds that it undermined the bedrock principle of free-market economics: that enterprise must succeed or fail on its own and without interference from government. Moreover, they said, fairness dictated that no business should be entitled to an advantage over competitors. The Business Roundtable, the National Association of Manufacturers, and Walter Wriston, chairman of Citibank and the nation's most prominent banker, denounced loan guarantees.

That Chrysler acknowledged past management mistakes was seen by lobbyists and public relations experts as vital to securing political support for its cause. Since Riccardo had planned to leave soon anyway, a bit of ritual scapegoating was staged to help the cause. Riccardo, who inherited most of Chrysler's problems from Townsend, resigned a few months earlier than anticipated, and on September 20, 1979, the board appointed Iacocca chairman and chief executive.

Iacocca stepped into the role he seemed born for, selling the rescue of Chrysler to an uncertain public. Speaking so eloquently that even the irascible Proxmire complimented him, Iacocca starred in a parade of witnesses who testified before House and Senate committees through the fall of 1979 as Congress debated whether and how to help Chrysler. An advance of tax refunds was rejected, and a relaxation of regulations didn't yield enough savings. Chrysler said it needed to borrow $1.2 billion but wouldn't be able to do so unless the government guaranteed the loan. The only acceptable aid, as far as Jimmy Carter and his advisers were concerned, was a loan guarantee.

Consensus-building had many faces, though the bottom line remained electoral politics. Coleman Young, mayor of Detroit, testified before Congress about Chrysler's importance as the city's single biggest private employer, bigger than GM or Ford. What he didn't say, but which everyone in Washington understood, was that Jimmy Carter was running for reelection in 1980 and the support of cities like Detroit was crucial. Moreover, Young wielded a great deal of clout with other big-city mayors. Douglas Fraser, president of the UAW, pleaded for quick approval of loan guarantees and promised that the workers were willing to "sacrifice." The union hadn't decided yet whether to endorse Carter or Ted Kennedy. Again and again, Chrysler's advocates asserted that conventional bankruptcy would destroy the company's chance to survive. Iacocca commanded the spotlight, however, by passionately describing the potential damage to cities, to tax revenue, and to national economic confidence if Chrysler went under.

Chrysler's economic arguments were clearly debatable. Many economists asserted that the automaker could survive a conventional bankruptcy in reasonable shape, assuming it used bankruptcy to reach more favorable agreements with the UAW, with suppliers, and with its bankers. Others pointed out that the economic slack from a Chrysler failure might have been taken up by General Motors and Ford—rather than Japanese imports—providing jobs and tax revenue to workers and hard-pressed communities.

Though many senators and House members may have believed

bankruptcy to be an equitable solution, they were reluctant to turn their back on a large, visible company, a household name with roots in so many communities, and one headed by a determined, personable chief executive—particularly since the loan guarantee probably would cost the taxpayers nothing. Before putting the vote to Congress, however, administration officials suggested that Chrysler begin persuading its suppliers, bankers, and workers to help reduce the automaker's expenses. It was clear from the rhetoric of many in Congress that without the concessions that were necessary in any bankruptcy, Chrysler didn't deserve their support.

In mid-October 1979, the United Auto Workers agreed to revise its contract with Chrysler, delaying wage increases during its three-year contract for a savings of $203 million. True, the labor contract was already $1.3 billion more expensive than the 1976–79 contract, but the contract concession, however small, was an unprecedented and significant step, since the union had never before deviated from its ironclad policy of charging all automakers the same wage rate. The union's policy of "pattern bargaining" was designed to prevent one automaker from gaining an advantage over a competitor by negotiating lower wages. Pattern bargaining had helped UAW members win the most lucrative pay and benefits in American manufacturing.

On December 20, 1979, the Chrysler Loan Guarantee Act passed both houses of Congress. Chrysler was eligible to receive as much as $1.5 billion in loan guarantees, provided it secured an additional $260 million in concessions from union workers (to be offset by $162.5 million worth of stock to be set aside for them), $125 million from salaried workers, and $1.4 billion from other constituencies like bankers and suppliers. A five-member government board that included Paul Volcker, chairman of the Fed, and Bill Miller, the Treasury Secretary, was empowered to supervise the Loan Guarantee Act. Chrysler was to receive nothing until the concessions were in hand.

"The hard part starts now," Iacocca said at the White House signing ceremony in early January of 1980. The union, convinced that Chrysler faced imminent collapse, quickly agreed to additional financial concessions. Automakers had long been in the habit of pleading poverty to workers during contract talks and were regularly disbelieved. This time, however, was different. Iacocca didn't have to use the hard sell. His approach was simple: first he reduced his own salary to one dollar a year (somehow a more credible description than "working for free"). He also invited Fraser to join Chrysler's board of directors, a radical step that horrified many executives in Detroit and shocked union members. No UAW member had ever served as a director of a company. Now, as an adviser to top management, Fraser was able to attend board meetings and examine for himself how strong or weak the company was and whether executives were telling the workers one thing and saying another among themselves.

Fortified by the additional credibility he built for himself, Iacocca was prepared to tell UAW negotiators, "I've got a shotgun at your head. I've got thousands of jobs available at seventeen dollars an hour. I've got none at twenty. So you better come to your senses." It was vintage Iacocca — powerful symbolism and distilled logic, punctuated by terse, clever prose. No pitchman had ever done better.

Several hundred banks around the world, which were owed about $4.5 billion by Chrysler, weren't nearly as impressed. Under the Loan Guarantee Act the banks were obligated to lend Chrysler another $650 million in order for the automaker to qualify. The chairmen of Citibank and Bank of America, as well as Pete Peterson, the head of Lehman Brothers, had gone on record suggesting that bankers take their chances in bankruptcy court and not go any deeper in debt to Chrysler. Only John McGillicuddy, chairman of Manufacturers Hanover Trust, Chrysler's lead bank, publicly argued that the banks should help keep Chrysler out of bankruptcy. But Manny Hanny's historic and unusually close relationship with Chrysler rendered McGillicuddy's support somewhat less persuasive.

While the union concessions took about two weeks to arrange, agreement from the banks turned into a six-month ordeal, led by Greenwald, Miller, and Salomon Brothers, Chrysler's investment bank. A new loan package needed unanimous agreement among the commercial banks, Greenwald and Miller decided, otherwise every bank would claim a special reason why it wanted to be paid and let off the hook. So Manny Hanny, with outstanding loans of $200 million, and small-town banks, with loans of less than $100,000, all had to agree.

Many banks initially thought as Citibank did, that they had a better chance of recovering some money in bankruptcy court. Their assessment was based on the questionable premise that money owed to them by the Chrysler Corporation (about $1.5 billion) was at risk while the more than $3 billion owed by its finance subsidiary, Chrysler Financial, was substantially secured by cars owned by dealers and retail customers and would be repaid promptly. Chrysler lawyers, however, informed the banks that in a liquidation all the loans probably would be regarded as Chrysler corporate obligations and, therefore, unsecured. Chrysler's viewpoint, confirmed by many of the banks' lawyers, cooled the bankers' enthusiasm for bankruptcy.

In mid-May 1980, after furious negotiations by Miller and Greenwald, the larger banks agreed in principle to restructure Chrysler's debt. In return Chrysler was willing to grant the bankers twelve million warrants to buy Chrysler stock at thirteen dollars a share. The warrants were good until 1990, meaning that when Chrysler got back on its feet and its share price rose above thirteen dollars, the warrant holders stood to profit. But since shares were then selling at $3.50, the thirteen-dollar strike price looked like a long shot. A few months later, the Chrysler Loan Board requested 14.4 million Chrysler warrants, on the premise that the American public also deserved a chance to profit if Chrysler prospered.

Through the end of May, only a handful of banks continued to resist. Over the next few weeks they were subjected to pressure from Chrysler executives and Treasury officials. As the number of holdouts grew smaller, the pressure on the remaining few increased since no bank wanted to be the villain who wrecked the restructuring. At the beginning of June, Chrysler was operating literally without cash and telling its 20,000 suppliers that a rescue was imminent. Many suppliers hadn't been paid since January, and any could have triggered an involuntary bankruptcy filing by insisting on payment. On June 19, 1980, the Treasury Secretary phoned the last holdout, the Twin City Bank in North Little Rock, Arkansas, to persuade the bank's officers to accept the restructuring. Twin City Bank's loan amounted to $78,000. Until they were reassured personally by Secretary Miller, the Twin City Bank's officers had refused to believe they were the reason for delaying the $1.5 billion loan guarantee.

The Loan Guarantee Act gave the government's five-member Loan Board statutory operating control over Chrysler. In the event of a failure, the government had first claim on most of the automaker's assets. Additionally, Chrysler was obliged to submit a yearly operating plan to the Loan Board and had to ask the board every time it wanted to spend more than $10 million—a fairly small sum in an industry where individual machines can cost that much—and Iacocca and other top executives were required to report monthly. Brian Freeman, head of the loan board's staff, had been appalled during the bailout of Conrail that the government had no control over Conrail's spending, however ill-advised. The government had no choice but to guarantee more lending. This time Freeman wanted controls so annoying that Chrysler was motivated to pay the loans back as soon as possible.

As G. William Miller told Robert Reich in a February 1984 interview for a book about the bailout, "Iacocca had to come to see me every month and report. They'd bring slides and things in just like a board-of-directors meeting. He hated it. He didn't want to report to me. I took it as if I were his chairman. . . . I told him, 'One thing I will do that will end up as a benefit to you is to make you hate the government so badly that the idea of getting them off your back will motivate you to make this a success.' "

A halter and bit in the hands of the government wasn't exactly what Iacocca had in mind when he climbed back into the saddle after his fight with Henry Ford.

THE SEEKERS

THE EARLY 1980s was a grim period for the auto industry. Chrysler's quest for government aid diverted attention away from the economic recession and the flood of cars imported from Japan. By the end of 1980 more than 250,000 Big Three workers were laid off; unemployment in Detroit had reached 14.6 percent, and in factory towns like Flint the jobless rate rose above 20 percent. The number of workers in the Detroit area employed by Chrysler dropped from about 80,000 a year earlier to 47,200 in 1980.

The Federal Reserve cooled inflation by ratcheting up interest rates, with the result that auto loans were costing as much as 20 percent. Gasoline prices were also rising in the turmoil following the 1979 overthrow of the shah of Iran and the oil embargo. As consumers postponed car purchases, the impact on companies and cities that depended on automaking was devastating.

On top of temporary shutdowns and layoffs caused by poor economic conditions, the Loan Guaranty Board demanded permanent cutbacks to shrink Chrysler into a smaller, more stable company. The terms of the loan guarantees required Iacocca to clear Chrysler's operating plans with the Treasury Department; Iacocca, to be sure, wasn't the sort of executive interested in running a "downsized" company. Bill Miller, the Treasury secretary, regularly bucked Iacocca's operating plans back to him, telling him to cut them more. "You haven't thrown off any ballast yet," Miller recalled telling Iacocca in a 1984 interview. "When the ship starts to sink, the first thing you do is get rid of ballast."

Chrysler's blue-collar workers, having foregone salary increases and benefits to ensure government help, suddenly realized that their sacrifices of pay and benefits didn't translate into job security. More than financial concessions were needed. Chrysler received $500 million in federally guaranteed loans in June and another $300 million in July, as the layoffs proceeded apace. Douglas Fraser, in his role as a Chrysler director, began to monitor plants targeted for shutdown so workers had a chance to propose efficiencies that might keep the plant open. For years manufacturing experts had been warning—and Fraser knew—that typical American auto factories employed far too many people and didn't utilize their labor efficiently. Publicly the United Auto Workers never conceded this point, but Fraser was well aware that factories could be run with far fewer workers if they and the managers cooperated.

When it came to saving time, effort, and material, the workers had as many creative ideas, if not more, than their bosses. Some involved little more than common sense—the better placement of machines, or better scheduling of parts delivery. Others took ingenuity, training, and some investment but yielded substantial savings. These ideas had rarely been proposed prior to Chrysler's emergency, because in their mutual antagonism union workers thought that plant managers just wanted to push them harder, while managers believed that workers were always looking for ways to goof off. In some measure, of course, they had read each other right.

This gap between toilers and thinkers was a factory-floor political division institutionalized by four decades of union contracts: workers correctly assumed that most supervisors didn't want to listen to their ideas about how to improve the plant. Besides, many workers wanted to put in their time, punch the timeclock, and go home without responsibilities or cares. In the face of Loan Board demands for more efficiency, cooperation by managers and workers was essential. Eliminating some of the more egregious instances of waste and featherbedding lowered costs to the extent that a few plants were saved from shutting down. As Samuel Johnson put it: "When a man knows he is to be hanged in a fortnight, it concentrates his mind wonderfully."

If Lee Iacocca was the figure who appeared before workers as the ax-wielding savage, it was Bill Miller and the Loan Guaranty Board who actually were ordering the cutting. His was no ideological mission; Miller wanted to make sure Chrysler repaid its loans so that the government wasn't stuck with $1 billion of bad debt. Miller knew how awful some corporate executives were at confronting reality unless they had to.

Chrysler, like other U.S. automakers, had never thought much about cultivating highly disciplined manufacturing practices. Detroit's creed, an outgrowth of its near-monopoly of the U.S. automotive market, was always to think big, to hang on to as much factory space as possible for the day when demand for cars surged, to worry less about the cost of waste or scrap than about ready access to parts and labor. To Detroit's way of thinking, it was a hundred times better to overinvest in new plants and parts inventory than to be caught without enough cars.

By the early 1980s, Iacocca and his fellow industrialists at GM and Ford possessed few reliable measures of how crude their manufacturing operations — not to mention the quality of their vehicles — had become as compared to Japanese competitors. This is not to say that Detroit executives had never visited or studied Japanese automotive plants. They had indeed observed Japan at close range. Each of the Big Three owned stakes in Japanese automakers. From their studies emerged the copious reports and memos that well-trained executives are expected to produce. But the practical effect of these observations was nil: no one pressed the issue too hard, lest colleagues suspect an unseemly admiration for things Japanese. After all, the Big Three still outsold everyone else by a mile. Detroit, ever imbued with the belief that it had little to learn from anyone else, saw Japan but failed to comprehend it.

The measuring sticks the auto industry needed in order to understand how far behind they had slipped didn't exist yet. A little-known market researcher in California and a team of academics from the Massachusetts Institute of Technology were about to change that.

Judging from his credentials, James David Power III was just the sort of bright young fellow the auto industry hired as fast as it could in the 1950s. He loved cars from the moment he learned how to drive the family's 1937 Chevy coupe in Worcester, Massachusetts. He was graduated from Holy Cross and had served as an officer in the Coast Guard. Trained as an M.B.A. at the Wharton School, his skill with numbers and statistics made him a natural for the cadre of finance men led by J. Edward Lundy, the legendary chief financial officer of the Ford Motor Company.

As it turned out, Power's love of cars couldn't overcome his distaste for corporate life. Ford might have held on to Dave Power anyway if the automaker had assigned him to study consumers and marketing, his true interests, instead of making him a financial analyst in the tractor division. Restless to learn about the proliferating uses of polling to discover why people like or dislike products and how companies use information about their customers, Power left to work as a market researcher for the McCann-Erickson advertising agency.

In the early 1960s, without a job, he left Detroit for the West Coast, where he and his wife started their own small market-research firm, conducting customer surveys for companies selling products as varied as chain saws and tape recorders.

Prospecting for new clients in 1968, Power visited the Los Angeles headquarters of a little-known Japanese automaker, the Toyota Motor Corporation. A poor company with scant experience in automaking, Toyota had imported its first car to California in 1958. In the subsequent decade Toyota tried, with mixed success, to expand its presence in the American market. In 1960, the automaker was forced to suspend exports to the United States because its cars were too flimsy and anemic for the expectations of American consumers. From this humiliating experience Toyota determined that it must study local conditions more rigorously if it hoped to build cars for people whose assumptions and habits were quite different from those of the Japanese. Toyota's studies paid off with the U.S. introduction in 1966 of a sturdy new Corona model, which came equipped with a powerful 1.9-liter engine and a series of options including air conditioning. At $1,860, the Corona was a reasonable alternative to the Volkswagen Beetle, which sold for $1,600. Toyota's exports to the United States grew to 39,000 vehicles in 1967 from 3,700 vehicles in 1964.

In 1968, Toyota hired Dave Power to dig deeper and find out what owners thought about their Toyotas, and, equally important, how customers were being treated by dealers. From 1964 to 1967, the number of Toyota dealers in the United States had tripled to more than 700. J.D. Power and Associates' first Customer Satisfaction Index (or CSI, as it would become known in industry jargon) was offered to Toyota as a rough measure of the problems customers encountered with their cars and the extent to which dealers resolved them to customers' satisfaction.

By the early 1970s, other Japanese automakers were buying Power's CSI surveys as well. Japanese executives were hungry for information, and they responded to it quickly. When complaints showed up in survey results, they were relayed to the factory or to dealers for resolution. Year by year, Power's CSI surveys grew more detailed while at the same time revealing more about customer preferences. In 1977, Honda agreed to spend the stupendous sum of $250,000 for a Power survey that analyzed every one of its 620 dealers. At a time when Detroit automakers were at odds with their dealers, fighting over payment of warranty claims or about who bore the greater responsibility for promoting sales, Japanese executives seemed curiously uninterested in apportioning blame, preferring to search for the root causes of problems and fix them.

Until the oil shocks of the 1970s Americans had chosen their cars mainly for their style, image, or because of an owner's slavish loyalty to a particular brand. A Buick was a doctor's car. A Mustang was for those who wanted to make a sporty statement. Owners of Chevy pickups almost invariably stayed with Chevy; likewise for owners of Ford pickups. In an era when society's notions about safety, the environment, and consumerism were evolving rapidly, very few sources spoke authoritatively about a car's inherent quality and reliability.

Naturally, Dave Power hoped the Japanese automakers' enthusiasm for his CSI surveys would pique the interest of General Motors,

Chrysler, and his former employer, Ford. Upon hearing his pitch, though, Detroit executives were coldly unimpressed: they believed their own consumer research to be the finest in the world. Surveying customer attitudes toward dealers was an idea that held promise, they told him, but there was no point measuring vehicle quality since everyone was receiving more or less the same level of quality. Obviously customers were happy since sales were strong. In any case, U.S. automakers didn't believe there was much that Dave Power could tell them about their customers they didn't already know. (The response to Power was eerily similar to the reception W. Edwards Deming got in Detroit decades earlier when he tried to teach statistical methods for the factory and the impact of management theory on product quality. The men of Detroit disregarded Deming even while Japanese industrialists were embracing his methods and continuously improving quality.)

The 1979 oil shock proved fortuitous for the Japanese automakers, whose vehicles were inherently more fuel efficient because of their low weight and compact size. Suddenly drivers were panicked— less over the high price of gasoline than over its very availability. Suddenly it *mattered* that the Toyota Corolla and other lightweight models could travel further and longer on a gallon of gas than the big Buicks and other cars built in Detroit.

Drivers who bought Toyotas, Hondas, and Datsuns primarily for their fuel efficiency were impressed by their utilitarian compactness and, in some cases, their high level of quality as compared to domestic models. The difference in quality wasn't always something you could put your finger on precisely, but an impression grew that the seams between metal parts seemed to be straight and even, that painted finishes had few defects, and that doors closed unusually snugly. The engines and transmissions lasted for what seemed like an extraordinary length of time. Moreover, Japanese designers had added thoughtful, pleasing touches like a cupholder or a remote switch to release the gas-tank lid. The difference wasn't just fewer things gone wrong, but things gone right.

Reports of high quality in Japanese cars were initially met with incredulity and contempt in Detroit. To some extent Detroit's memories were fixed on the little Honda 600s and Subaru 360s of the early 1970s, minicars that could barely get out of their own way and weren't safe in a collision. The newer models, late 1970s versions of the Toyota Corolla and the Honda Civic, were estimable, even charming, vehicles. But to many who bought big Chevies and Fords — and certainly to those who sold them — the Japanese cars were nothing but "shitboxes," the automotive equivalents of Cracker Jack toys. Of course, people who had owned Hondas and Toyotas bore witness that Japanese cars were quite worthy. Reviews in *Consumer Reports* and automotive enthusiast magazines like *Car and Driver* had begun to describe some Japanese models respectfully. But no magazine was strong enough at that time to contradict Detroit's advertising and marketing onslaught. An independent, impartial measure of automotive quality aimed at consumers didn't yet exist.

Armed with new software, and willing to bear the high cost of computer research himself, Power conducted his first mass survey of car owners in 1981. Using registrations of new cars to randomly compile tens of thousands of names and addresses, Power mailed out questionnaires asking how many defects drivers discovered in their cars during the first months of ownership and how willingly dealers had handled the repairs. The number of questionnaires returned yielded statistically significant results for every make of car.

The results of Power's survey, confirming anecdotal impressions of people who owned Japanese and American cars, didn't surprise him. They showed that Japanese franchises, on average, had far happier, more satisfied customers than General Motors, Ford, or Chrysler. Japanese cars, particularly Toyota and Honda, were well built; and their dealers for the most part treated customers fairly and with respect.

Auto industry research specialists immediately disputed Power's methods. Ford, for example, advertised heavily with the slogan "Quality Is Job One," in rebuttal to Ford's poor score in Power's CSI studies. Ford research specialists had little choice but to argue that Power's methods weren't valid. Power paid independent statistical experts to examine and report on his surveys' validity. Their reports showed his surveys to be not only valid but, in some cases, startlingly accurate. One J.D. Power survey uncovered a defect in Mazda's rotary engines before the automaker's management realized there were enough complaints from owners to suggest a chronic problem.

The J.D. Power survey steadily gained credibility among manufacturers. Dave Power was amassing data on every brand, measuring how complaints varied from year to year or among models built in more than one assembly plant. The results were dramatic. Some Japanese brands had fewer than half the complaints of many domestic brands. Now executives of the domestic automakers were able to judge for themselves what randomly selected customers were saying about their cars in comparison to what owners of Japanese cars were saying. Toward the end of the 1980s, Power was mailing out four to five million questionnaires annually. The questionnaires were six to eight pages in length, contained about ninety questions, and took about half an hour to fill out. J.D. Power included a one-dollar bill in each envelope, which helped to elicit a phenomenal 40 percent response rate. Americans, as enthusiastic owners of cars, were eager to describe their automotive joys and disappointments.

The big break for J.D. Power and Associates came unexpectedly. In January 1985, Power was at home watching the Super Bowl game when a Subaru commercial was broadcast boasting that Subaru had been rated "number two, behind Mercedes, in the J.D. Power and Associates Customer Satisfaction Index." Power was surprised. He knew Subaru's CSI was second only to Mercedes, but he had no idea the automaker intended to tout its standing in advertisements.

Subaru's advertising featuring its J.D. Power rating turned out to be an important factor in the Japanese brand's popularity during the mid-1980s. Soon other Japanese brands, including Toyota and Honda, were also advertising their J.D. Power ratings. Shoppers who paid attention to the advertisements began asking dealers about a model's J.D. Power rating, just as they might ask about a gasoline's octane rating or how many stars a restaurant received in the Michelin Guide. Honda created a side-window decal for its Acura Legend luxury sedan attesting to its number-one CSI rating.

If customers felt wary about buying a Japanese car—thinking it unpatriotic to patronize a Japanese company or detrimental to the

American economy to buy a car not built in Detroit — they now had a high J.D. Power rating to provide a counterweight to that wariness. The bottom line for many, if not most, American shoppers was a deep belief in value, the belief that one was entitled to receive the best merchandise for the money.

By the mid-1980s the domestic automakers saw that J.D. Power was having a dramatic impact in dealer showrooms. The CSI survey results reinforced the good word of mouth Japanese automakers were getting. And sales results bore witness to the surveys, as Japan's share of the U.S. automotive market crept higher and higher. Finally GM and Chrysler began purchasing Dave Power's surveys and research data; Ford held off until 1989 but finally relented.

Initially domestic manufacturers bought J.D. Power surveys primarily to identify which of their dealers were receiving the most complaints from customers. (In the view of most auto executives, improper treatment of customers, not poor quality, was the main reason for dissatisfaction with a car.) Quite often the automaker threatened sanctions against dealers with a low CSI rating; a dealer with a low CSI rating might find his application for a new franchise turned down. Power, like Deming, was convinced that threatening dealers was counterproductive and actually dampened their motivation for improving customer service. "This will set you back a half dozen years," he pleaded with executives in Detroit. "The surveys were never meant to be used punitively."

Power's words were prophetic. Many dealers, instead of trying to improve their service so that customers were treated properly, found unethical methods for boosting their CSI ratings. Some tried to stuff the ballot box with fake questionnaires, a scam that Power eventually foiled with encoding. Some dealers mailed out offers of free gifts to customers who brought a J.D. Power questionnaire to the dealership so a salesperson could "help" fill in the answers.

Through his surveys Power discovered that automakers hurt themselves more by failing to correct defects than by selling defective cars. If automakers provided incentives for dealers to help customers, like paying warranty claims without hassle, customers tended to forgive minor defects. By carefully analyzing the results of surveys,

Power was able to measure the average rate of defects, including everything from a small scratch on the hood to an engine that stalled. The surveys isolated how a customer was treated by the salesman and the service department, and whether the warranty was honored if repairs were needed. If vehicles came off the assembly line with poor manufacturing quality, the J.D. Power rating caught it; if automakers were rejecting warranty claims, the ire of customers was detected quickly. From the beginning, the Big Three scored poorly against Japanese competition on J.D. Power surveys, recording five and six times the number of complaints as Japanese nameplates.

Since denial hadn't worked, the Big Three had little choice but to improve quality, a process that took several years. In 1989, the Buick division of General Motors enjoyed the payoff for improved quality in precisely the manner Dave Power envisioned when he concocted his surveys. In a J.D. Power Initial Quality Survey of 1989 models—a measure of owner-reported problems with the car after ninety days of ownership—the Buick LeSabre in one year leaped from forty-fourth to second place (behind the first-place Nissan Maxima). LeSabre's improvement was partly due to a campaign of meticulous attention to quality at GM's Buick City assembly plant in Flint, Michigan, where the model was built. When Buick advertised its improvement in the survey, sales of LeSabre instantly doubled.

Detroit found itself in a pickle. Over a period of a half-dozen years, J.D. Power and Associates had evolved from near obscurity to a position of vast influence. Even buyers who had never heard of J.D. Power were being swept up in the migration to Japanese brands that was fanned by shoppers—and automotive journalists—who heeded the survey closely. The automotive public was paying careful attention to Dave Power's unbiased measures of quality and service; and the measures were not flattering to General Motors, Ford, and Chrysler. If the Big Three automakers intended to win back customers, there was little choice but to score higher with J.D. Power. To do that, American automakers had to clean up their act.

Today, with gasoline cheaper and more widely available than at any time in recent history, it's easy to forget that during the direst moments of the 1979 energy crisis the practicality and future of the internal-combustion engine as a means of powering passenger cars were being called into question.

The U.S. auto industry had been under siege for almost a decade as regulators demanded vast improvements in safety, environmental cleanliness, and fuel efficiency. The long queues to buy gasoline in 1979 appeared as a sign that American society should finally rethink the automobile's long-term viability. To the extent that the questions were addressed rigorously and objectively, the answers obviously held implications for industry, labor, government — indeed, the entire modern world, where some 400 million motorized conveyances were already operating.

The need to thoroughly reassess motorized transportation represented the viewpoint of a group of academics in Cambridge, Massachusetts, led by Dan Roos, who was director of transportation studies at MIT's School of Engineering, and Alan Altshuler, a professor of political science at MIT. Altshuler had gained some local renown as the Massachusetts secretary of transportation who halted a twenty-year effort to superimpose a grid of freeways over Boston, which critics said would ruin the historic flavor of the city. Assisted by Jim Womack, a graduate student in political science, Altshuler wrote *The Urban Transportation System* (1978), a book that attempted to catalogue the ills of the automobile-dominated transportation system and examine solutions, from building more subways to imposing high use taxes on automobiles. Roos, whose academic specialty was transportation, was deft at raising money for large-scale policy-oriented academic projects. Roos and Altshuler believed that a comprehensive study of the future of the automobile was timely and would surely attract sponsors, and that the results might help shape society's thinking about how to use transportation technology.

In a historical sense, MIT was a fitting location for such a study, as it was there that the legendary Alfred P. Sloan, Jr., had studied electrical engineering before getting into the automotive-parts business. Ultimately, Sloan sold his business to GM, where he became an executive and organized the troubled company in the 1920s into the world's mightiest automaker. As chairman of General Motors, Sloan contributed generously to MIT, where his money was used to endow the Sloan School of Management. General Motors maintained ties to MIT after Sloan's death, funding studies of automotive safety and other technical subjects.

Roos and Altshuler barely scraped together the money to support a comprehensive study of the automobile. The automakers were less than enthusiastic about the project, since they believed that scrutinizing the basic assumptions of automobility was, more or less, a silly waste of time. If MIT's administration was worried that the researchers might come up with something powerful enough to offend GM, a major contributor to the university, it needn't have been. The group finally invited to study the automobile's future was an international mélange of social scientists, professors, politicians, and technocrats pursuing narrow academic and national interests. During their international conferences, the members from Japan and Sweden gave long-winded paeans to free trade—in this case, the desirability of open markets for their cars. The French members, on the other hand, fulminated about what they called "free-trade imperialism," asserting that aggressive automotive exports by Japan and others threatened to undermine their national sovereignty. Several of the professors delivered abstruse, theoretical orations, complete with charts, slides, and impenetrable algorithms.

Those who might have contributed greatly, or at least have provided a fresh perspective, the executives of the big automakers, mostly ignored the meetings. General Motors and Ford each assigned fairly senior public relations executives to keep an eye on what was going on. Chrysler, claiming it had no one to spare since the company was close to bankruptcy, sent no one. Only the Japanese automakers, always hungry for new information and eager to better comprehend the environments in which their industry was operating, sent high-ranking officials.

In 1984 Altshuler, Roos, Martin Anderson, and Daniel Jones published *The Future of the Automobile,* a collection of the group's findings. Essentially the book restated in greater detail the headaches of depleting resources, trade tensions, pollution, crowding, competition, and labor unrest that had prompted the study. It was a harmless, ineffectual document that spotlighted some significant trends, such as the growing cooperation between automakers to reduce the financial risk of capital-intensive projects. Of course many technical challenges loomed, the book concluded, like the demand for cleaner tailpipe emissions and better safety equipment. But technologies also existed to provide solutions, and the automakers were hard at work perfecting airbags, catalytic converters, batteries for electric cars, and so forth.

The Big Three might have profited from one very prescient observation buried in the study: that the American and European auto industries were not as productive as Japan's. The book included one strikingly memorable graph. Using a variety of data for its estimates, the graph documented the number of total hours of employment required to produce a single motor vehicle in the United States, Germany, and Japan for the years 1970 through 1981. In 1970, according to the graph, an American automaker expended about 200 hours of labor per vehicle, including hours devoted to final assembly, parts manufacture, and management; eleven years later the total hours had risen a bit, to about 210 hours. In Germany, by comparison, the number of hours of labor per vehicle was still slightly under 200 in 1981. The productivity improvement in Japan, meanwhile, had been stunning: whereas in 1970 the Japanese automakers needed an average 255 hours of labor to produce a car, by 1981 the industry needed only 140 hours, a 45 percent improvement.

Without a doubt, during the 1970s the Japanese auto industry underwent a stunning leap in sophistication — so much so that it was operating at a far higher level of efficiency than the two most experienced and prosperous auto industries in the world. This, then, was the group's most important, albeit little-noticed, finding, which at least helped to explain why Japanese vehicles were so much cheaper than U.S. models. By the fall of 1984, when the results of the study were published, Toyota, Honda, and Nissan clearly constituted a force in automaking for reasons besides just efficiency. The second-generation Honda Accord sedan, in particular, had become the choice of smart Yuppies who were interested in good value and trouble-free driving. The secrets of exactly how the Japanese manufacturers were making their cars durable and appealing was not well understood in Detroit. But the graph of relative efficiency among American, German, and Japanese industries raised the intriguing possibility that Japan's factories and engineering laboratories—indeed, their entire approach to mass manufacturing—differed radically from the orthodoxies of Flint and Cologne.

In 1984, when the MIT research team released its findings, American hostility toward Japan was on the rise, inflamed by the tide of imported cars and the resultant job loss in factories here. Just a few years earlier, the Reagan administration had concluded an informal agreement with Japan to limit the number of cars exported to the United States. But now Honda and Nissan were operating U.S. assembly plants that produced cars which were not governed by the agreement; and although these factories employed American workers, many of the cars' components were imported from Japan, threatening jobs in the U.S. automotive parts industry. So the auto industry, led by Lee Iacocca at Chrysler, was calling for new, stronger trade protection.

Unfortunately, the MIT researchers had not focused on the essence of Japanese automaking expertise and could offer no prescription for change. For if they had, then the U.S. automakers could at least have studied what they needed to achieve parity. The U.S. economy was strong in the mid-1980s, and auto sales were still rising. When sales inevitably slumped during the next downturn, auto industry executives were sure to argue once again for trade barriers against the Japanese. That step, the researchers knew, would be foolish, for it would only prolong the industry's denial, it would only delay the inevitable Judgment Day. What the U.S. industry needed was a detailed study of Japanese manufacturing, exposed in greater detail and with more supporting data—a study so incisive and clear that it would provide a roadmap for policy-makers, factory managers, mayors, and practically anyone with an interest in keeping American carmaking alive.

Someone else was having similar thoughts, a clever Korean-American named Jay W. Chai. Chai, then executive vice president of the U.S. subsidiary of the Japanese trading concern of C. Itoh and

Company, had been invited to the MIT automotive research conferences because of his expertise in international business and his
knowledge of the Japanese automobile industry. Having served as a
director of Isuzu Motors and having visited countless Japanese automakers, he was somewhat disappointed by the MIT study.

"You have only begun to peel away the first layer," Chai informed
Roos. As one of the very few ethnic Koreans to attain a high post in a
Japanese company, and having studied business at the University
of Southern California, Chai claimed an unusual, perhaps unique,
perspective on how Japanese automakers operated. Chai possessed
another unique qualification: he was married to Maryann Keller, the
highly regarded Wall Street automotive analyst, who chatted regularly
with the industry's most important figures. A thoughtful and engaging
man who moved discreetly in the circles of chief executive officers,
bankers, and directors, Chai had come to be regarded as "the invisible man" of the world auto industry.

"You must try again but this time go much deeper," Chai told
Roos. "If you need money I will help you find it." Years earlier he had
helped General Motors buy an equity stake in Isuzu. He was friendly
with Roger Smith, GM's chairman, and realized that Smith had accepted the fact there was no choice but to learn what the Japanese
automakers were doing. Indeed, Chai had acted as a go-between for
General Motors and Toyota in a joint carmaking venture in Fremont,
California, which was meant to help GM understand Toyota's production system. Chai knew that the Japanese automakers felt enormous
political pressure to provide their American counterparts with technical assistance. With Chai's recommendation to automakers around
the world, raising $5 million for another MIT study wouldn't be too
difficult.

Thus was born in 1985 the International Motor Vehicle Program
at MIT, under the direction of Dan Roos, Dan Jones, and Jim Womack.
This time the project wasn't relying on the esoteric theories and con-
voluted equations of academics and policy-makers. Instead, MIT
hired research specialists with specific knowledge about how business and industry really operated in the United States, Europe, and
Japan. Specialists were hired who understood factories and machines;

others were familiar with automotive parts suppliers and retail dealerships; still others knew about automotive technologies and the intricacies of product design and marketing. The team determined to study the major automakers and factories around the world, to ask industry executives and government policy-makers what they thought about the findings, and to compile the results in a readable form. After preliminary research, Jim Womack and others already hypothesized that the Japanese automakers, and Toyota in particular, set the standard for excellence. Their 1986 comparisons of factory productivity and quality at General Motors, Toyota, and the GM-Toyota joint venture in Fremont reinforced their hypotheses. The researchers were pretty sure the top brass at the automakers, especially the Americans, would be too nervous to cooperate by opening their factories. Hence, Womack and John Krafcik, a factory expert who had worked at the GM-Toyota factory, drove over to GM's Framingham, Massachusetts, assembly plant and bluffed their way into a guided tour, complete with briefings that allowed them to calculate productivity. Their motto became: "Don't ask." And it worked.

The Framingham plant proved to be one of the most antiquated and inefficient in GM's network. But the visit was revealing for another reason: when researchers confronted the plant manager with their data, they were shocked at how little appreciation the man had about what General Motors was up against in terms of international competition. "I could match Japanese productivity if I was building tinny little cars," the GM plant manager thundered when shown the researchers' statistics, "but I'm building Oldsmobiles!" To the extent that this plant manager was typical, the researchers concluded, American automaking was in deep trouble.

John Krafcik called on his former *kachōs*—his Japanese bosses—for whom he had worked in California to arrange similar surreptitious tours of the GM-Toyota joint-venture plant and of Toyota's Takaoka plant. The results of the surveys were presented in the fall of 1986 at MIT's Endicott House conference center. This time many automotive executives, middle managers, and other industry experts from around the world were present to hear how badly American automaking was lagging behind the standard set by Toyota.

Participants at the conference presumed that Womack and Krafcik had enjoyed the cooperation of GM and Toyota, and as a result the researchers found it much simpler to gain permission to visit the rest of the world's automakers, on condition that the identities of individual plants were concealed when the results were published.

The results from the surveys of factory productivity embarrassed some of the executives from sponsoring automakers. To others the results, though embarrassing, represented an opportunity. Finally they had impartial evidence of a documented gap in productivity with Japan that could be used to prove to hidebound managers and the United Automobile Workers Union just how serious were their problems.

We've won the war, Womack thought; the auto industry will embrace our findings as serious and credible. Just as Dave Power had invented a way for consumers to measure and judge quality, MIT had taken the first step in shattering the denial of those who ran American auto plants.

PERILOUS PROSPERITY

DETROIT HAS ALWAYS been a feast or famine town. Either factories were shutting down, workers drawing unemployment, and automakers pushing surplus vehicles onto reluctant dealers, or assembly lines were humming, factories were hiring anyone who breathed, and dealers were begging for more cars. When consumers felt prosperous and secure about their jobs, they bought cars and the national economy glowed with good health. In a vibrant economy, consumers flocked to showrooms without the enticement of discounts or promotions.

The resulting sales unleashed a torrent of cash that poured into automakers' coffers with the blunt force of a tidal wave. Cash flow from healthy automotive sales exceeded what automakers needed to declare dividends, repay loans, parcel out bonuses, fatten research budgets, repair the roof, and buy everyone a new desk. As long as sales held up, delighted executives were literally at a loss as to what to do with the prodigious sums of money that glutted their treasuries. The next economic downturn was always just around the corner, and the cash tap could slam shut just as quickly as it opened to full flow. Still, intoxicating moments of prosperity and self-satisfaction were rare and to be savored.

When the recession of 1980–82 gave way to a strong economic recovery, the cash spigot opened wide for the Big Three. Under the leadership of chairman Roger B. Smith, the General Motors Corporation embarked on the biggest acquisition spree ever by a public corporation. General Motors bought the Electronic Data Systems Corporation in 1984 for $2.5 billion; the following year GM bought Hughes Aircraft for more than $5 billion. Smith spent billions more on fancy robotic gadgetry (much of which later was junked as impractical) and a slew of smaller companies, including several financial institutions. The Ford Motor Company, which at one point amassed $10 billion in cash, had bid unsuccessfully for Hughes. Subsequently Ford acquired First Nationwide, the nation's biggest chain of savings-and-loans, for $493 million; the Associates, a consumer lending and financing company, for $3.4 billion; and Hertz, the rental-car company, for about $1.3 billion. General Motors and Ford reckoned that nonautomotive subsidiaries could generate profit during periods of weak automotive demand, to smoothe the cycles of boom and bust.

Automakers also used excess cash to repurchase their own stock from shareholders; this maneuver was thought to create value because it increased per-share profits for remaining shareholders, thereby making the stock more attractive and, executives hoped, causing its price to rise.

Another truism of Detroit: the U.S. auto industry takes its lead from General Motors. Starting in the 1930s "the General," by virtue of its commanding size and clout, set the auto industry's agenda in styling, pricing, technology, compensation, labor relations, and in just about every area that required executive judgment. The price of a Buick sedan, for example, established the price ceiling for all luxurious family sedans. If General Motors decided to increase production of convertibles, every automaker immediately considered doing the same. In 1983, when Roger Smith decided that General Motors would use its cash to diversify outside the automobile business, Smith's actions were seen by crosstown rivals as an endorsement of nonautomotive diversification.

That Lee Iacocca began to wonder about how to spend surplus cash four years after Chrysler's near bankruptcy testified to the dramatic impact of economic recovery on an automaker's finances. Car buying was strong, and overnight Iacocca was riding tall again. Chrysler's K cars, fuel-efficient compacts selling as the Dodge Aries and Plymouth Reliant, had succeeded in attracting consumers who remembered the gasoline lines and were convinced by images of Iacocca on television pledging, "If you can find a better car, buy it."

By mid-1982, just eighteen months after drawing its last installment of government-guaranteed loans, Chrysler was earning an operating profit. In the first quarter of 1983 Chrysler posted record earnings of $172.1 million. With cash flowing in so quickly, Chrysler decided to repay $400 million of the $1.2 billion it owed in guaranteed loans — eight years early. In July of the same year, 1983, a beaming Iacocca chose the National Press Club as the forum to announce that Chrysler was repaying the remaining $800 million of loans by year's end. "We believe this action should tell the financial community, but especially the American car buyer, that we are here to stay and have the strength to compete with anybody in the world," he told reporters.

Iacocca had more than one reason to repay the debt early. Interest rates were falling. New lower-interest debt would be less expensive than the guaranteed loan. Moreover, until guaranteed loans were repaid and dividends were restored to holders of preferred stock, Chrysler remained subject to the supervision of the Loan Guaranty Board. Supervision entailed clearance from Treasury Department officials before every significant expenditure, humiliating reviews every month in Washington, no executive bonuses, and no executive perks like corporate jets.

Since Chrysler no longer needed government guarantees to borrow money, Iacocca was eager to shed the Reagan administration's yoke. Iacocca detested the dense atmosphere of Republican privilege in Washington. The Republicans reminded him of the WASPs who had mocked him as a youngster in Allentown, and of the patrician Henry Ford, who had thwarted his ambitions at Ford. Iacocca wasn't affiliated with a political party, but he sympathized with the Democrats, the traditional choice of oppressed ethnic groups and underdogs.

If Iacocca resented the Republicans, the feeling for him in Washington was certainly mutual. The Reagan administration inherited the

Loan Guarantee Act from President Jimmy Carter and was obligated
to honor it. But it did so with little enthusiasm. Supply-side economists and ideologues in the administration, who saw no reason to
protect the American auto industry from Japan, made no secret of
their disdain for Iacocca or for "industrial welfare," as some referred
to Chrysler's loan guarantees. The Reaganites thought Iacocca nervy
for attacking government economic policy even as Chrysler benefited
from Washington's help.

Donald T. Regan, the former chairman of Merrill Lynch who was
Reagan's Treasury secretary, held a particularly dim view of Iacocca
and his bluster. Iacocca had lobbied everyone in sight to revoke the
provision of the Loan Guarantee Act requiring the sale of Chrysler's
corporate jets. Now that Chrysler was enjoying the fruits of a healthy
economy, Iacocca was again trying to change the terms of the
agreement, asking the government to forego its 14.4 million warrants
for Chrysler stock, granted under terms of the Loan Guarantee Act.
The warrants gave the U.S. government the right to buy Chrysler stock
at thirteen dollars a share. At the time the warrants were actually
granted Chrysler stock was selling for about five dollars a share and
no one thought there was much likelihood that they would increase in
value. But Chrysler stock, propelled by its recent profitability, was
now selling for about thirty dollars a share, making the government's
warrants worth nearly $250 million.

In the opinion of Regan and other Treasury officials, Chrysler's
request was presumptuous. Why on earth should the government give
up warrants worth $250 million? G. William Miller, the former Trea-
sury secretary who negotiated the bailout under Jimmy Carter, wrote
as a private citizen urging members of Congress to oppose cancella-
tion of the warrants.

Iacocca was outraged. He argued that the loan guarantees actually had never been much of a risk, since the government held a lien
on Chrysler's assets. He accused the administration of acting like a
loan shark and trying to extract disproportionate payment. The government's refusal, he told reporters, "doesn't meet the test of fairness,
or even decency. You can argue about the technicalities, but the oper-
ative line is that when you pay back a loan of that size seven years

ahead of schedule, you often renegotiate the terms because there is
no risk left."

Exercising the warrants didn't cost Chrysler any money out of
pocket. But Chrysler didn't want new shares diluting existing shares.
The automaker offered to buy the warrants back from the government
for $211 million. Again the answer was no. Instead, Regan ordered
an auction, knowing that several Wall Street investment firms were
willing to speculate on the future appreciation of the warrants. Iacocca and Regan traded potshots in the newspapers, Iacocca portraying Regan as Shylock, and Regan accusing Iacocca of misleading
the public.

Chrysler's chairman didn't grasp that, politically as well as ideologically, the administration was not about to forego $250 million,
possibly more, money that taxpayers regarded as their legitimate reward for bearing the risk of the bailout. Iacocca felt thwarted and took
Regan's refusal personally. "I've dealt with an administration and
people representing that administration who ideologically didn't like
our deal," Iacocca pouted, "and I've gotten the feeling that they just
wish I'd go away and maybe fail."

Iacocca tried to lobby Ronald Reagan personally. George Bush,
a casual acquaintance of Iacocca's, arranged a twenty-minute audience. On the appointed day the Great Communicator only wanted to
talk about Hollywood and the techniques used in filming Iacocca's
commercials for Chrysler. When the meeting was nearly over, Iacocca
tried to squeeze in an argument for cancelling the government's warrants.

But Reagan raised his hands: "You'll have to take that up with
Don Regan." Iacocca later told Chrysler executives on his return to
Highland Park that Reagan was either "very dumb or a lot smarter
than he seemed."

In an auction held the following month, Chrysler entered a winning bid of $311.1 million for the warrants, about $15 million more
than the next highest bidder—a hefty expenditure for a company
that just a few years earlier had trouble meeting its payroll. However, Iacocca and Chrysler's top executives were betting heavily on
the future appreciation of the automaker's stock price. Stock and

stock options constituted the backbone of Chrysler's executive compensation. By holding down the number of shares outstanding, pay packages would be worth more. In 1979 Iacocca had drawn a much-publicized salary of one dollar, symbolic of his concern for Chrysler's dicey situation. What wasn't well known was that Iacocca's $360,000 missed annual salary would be delivered to him two years later in the form of stock, based on its price appreciation. The value of Chrysler stock dropped during that period, so Iacocca missed collecting the salary portion of his pay package. Stock option grants he received during the period he was drawing his one-dollar salary had no immediate value, but they stood to appreciate tremendously if Iacocca managed to drive Chrysler's stock price higher.

Iacocca's first payday following repayment of the guaranteed loans in 1983 was a gala financial event, the antithesis of the austerity of the preceding four years. Since the death of his wife, Mary, in the spring of 1983, rumors had swirled that Iacocca was going to retire once the loans were repaid. He did little to quell the rumors, realizing that, in one sense, the rumors made him seem more attractive and valuable to the Chrysler board. Indeed, by creating, that summer, an "office of the chief executive"—consisting of himself, Gerald Greenwald, Ben Bidwell, and Hal Sperlich—Iacocca intensified speculation on Wall Street and in Detroit that he was positioning top executives as part of a succession plan.

Chrysler directors were eager for Iacocca to stay. They had been dazzled by his performance before Congress and in Chrysler's advertisements. The automaker now had momentum, which a change in leadership could only threaten. Chrysler's directors were willing to stretch to keep Iacocca from retiring, but it wasn't going to be cheap. His terms: 150,000 Chrysler shares, worth $4.2 million, if he stayed through 1986; another 50,000 shares, worth $1.4 million, for staying through 1987. New options to buy 300,000 shares at twenty-eight dollars (on top of options to buy 902,000 Chrysler shares he already had been granted at prices from $6.25 and $30.38). And, of course, $475,308 in salary for 1983.

The munificence of Chrysler's executives and directors did not escape the notice of the United Auto Workers Union. The union valued Iacocca's stock options, based on normal appreciation, at about $14.4 million. Douglas Fraser, the UAW's outgoing president, had accepted a seat on Chrysler's board during the bailout, so he and the union were well informed of the automaker's rapidly improving finances. To the union, improving finances presented the opportunity to restore the workers' wage and benefit concessions, valued at $1.1 billion. Without the approval of concessions by Chrysler workers in 1979 and 1980, the government never would have approved the loan guarantees.

Although UAW leaders were offended by the pay packages of Iacocca and other auto executives, they also believed that Chrysler remained vulnerable in the event of a sudden downturn in the economy. In light of rising sales of Japanese cars, Fraser understood that pressing too hard for higher pay and benefits could kill the goose. Yet each time Fraser tried to explain this rationale at union headquarters, Iacocca made yet another pronouncement about Chrysler's miraculous revitalization. Fraser was trying to lower the expectations of union rank and file while Iacocca kept trying to bull the stock. After a rosy financial forecast by Iacocca during labor negotiations in late 1982, Fraser bluntly told the *New York Times:* "If he would just keep his mouth shut, it would greatly simplify our lives. They're still a weak company, financially speaking."

Iacocca realized he had no choice but to restore much of the pay and benefits Chrysler workers had given up, but he also wanted to reduce cost-of-living allowances and health-care benefits to keep overall labor costs down. Despite the union's strategy to restore pay parity among Big Three factory workers, Iacocca was determined to keep Chrysler's labor costs below those of General Motors and Ford. But that strategy flew in the face of the union's commitment to pay parity among the Big Three. In 1982 the union struck Chrysler for five weeks before coming to contract terms. The following year, Owen Bieber, the new UAW president, managed to negotiate a new contract without a national strike.

But Chrysler's relations with the union remained tense. Many workers felt betrayed because they believed Iacocca had conveniently forgotten their sacrifice. He was raking in millions while they were being told the company couldn't afford to pay them what workers were getting at Ford and General Motors. But dissatisfaction over pay and benefits were only part of their bitterness: workers at a Chrysler factory in Twinsburg, Ohio, that supplied metal stampings for most of Chrysler's assembly plants struck for five days in late 1983, complaining of forced overtime, fatigue, and unsafe working conditions. A few weeks earlier a worker had been crushed to death in a stamping press; the company said the worker failed to follow safety procedures.

Lee Iacocca was receiving a great deal of public adulation for saving Chrysler. The union and workers didn't begrudge Iacocca the limelight, for they recognized that his popularity helped sell cars and, thus, kept them on the job. Executives were paid executive salaries. But workers believed that they had played a crucial role as well. The spirit of shared destiny among those with a stake in the automaker's survival had been the cement that held everything together.

The sense that shared destiny was crumbling, that Chrysler was returning to business as usual, reminded workers of the old maxim that union leaders had tirelessly preached to them: Trust was fine as an abstract concept, but nothing beat a tough contract for ensuring fair terms of employment.

The success of Chrysler's front-wheel-drive K cars solidified Hal Sperlich's already considerable reputation as one of Detroit's most imaginative product engineers. Sperlich had played a key role in the development of the Mustang, which had established Iacocca's reputation. Henry Ford II, in one of his fits of temper, had fired Sperlich in the mid-1970s, but industry cognoscenti suspected that his firing also was really an attempt by Henry to hamstring Iacocca, whom by that time he had grown to despise.

When Sperlich was hired at Chrysler in 1977 he went to work immediately on the K car "platform," as a basic vehicle chassis and body configuration is called. Sperlich also revived a pet project that had languished at Ford, the mini-max, sometimes called the minivan. The minivan was an idea that took the best attributes from a fuel-efficient station wagon and joined them with the spaciousness and utility of a van. By using the front-wheel-drive configuration of the K car, the minivan could be designed low to the ground, because it didn't have a long drive shaft. Women in particular seemed favorably disposed to it since it could carry kids and groceries yet wasn't too big or intimidating. A woman wearing a skirt or dress didn't like to climb up or down from a tall vehicle. Men liked the fact that, unlike normal-size vans, a minivan fit into the garage.

Extensive research at Ford had rated the minivan as likely to succeed, but the project was going to be expensive for Ford initially since it required the development of a new front-wheel-drive platform. Due to the expense of initial capital outlays—and a lack of enthusiasm by Henry Ford—Ford executives repeatedly shelved the mini-max. Besides, in the back of their minds, Ford executives wondered why, if the minivan was so hot an idea, General Motors hadn't done it first.

After Iacocca arrived at Chrysler in 1979, Sperlich immediately proposed building the minivan that Ford didn't want. Most of the conceptual thinking had been done. The K platform that had been used for the Dodge Aries and Plymouth Reliant would work perfectly for a minivan. The plant capacity was available. All that remained was a body design and detailed engineering specifications. Of course, as a new *category* of vehicle, the odds against the minivan were far greater than for a typical new model. But it was a gamble with a psychic as well as a financial payoff: Ford had rejected the minivan; if Sperlich and Iacocca were right, they could savor the joy of shoving their success in Henry Ford's face.

From their first day in showrooms, in January of 1984, Chrysler's minivans were a hit beyond all expectations. Capable of seating seven adults, the minivan could also be ordered without seating for passengers and used as a light delivery vehicle for florists or other tradespeople. Minivans sold briskly at prices ranging from $8,000 to $11,000 fully equipped. Since they were regarded as trucks for the purpose of federal safety standards, Chrysler was initially able to avoid all sorts of safety equipment, such as headrests, that were required for cars. Automotive analysts quickly grasped the minivan's salutary impact on Chrysler's balance sheet. The vehicles were yielding as much as $5,000 in gross profit per unit, an astonishing margin. At 200,000 units annually, minivans promised a potential gross profit of $1 billion for Chrysler.

Iacocca immediately recognized the need for more production capacity. Overruling Sperlich and others, he ordered a second assembly plant in St. Louis readied to produce the minivan, in addition to the existing minivan assembly plant in Windsor, Ontario.

Best of all, from Chrysler's standpoint, Ford and GM were still in denial regarding the minivan's popularity. The automakers insisted that their rear-wheel-drive vans — GM's Astro and Safari, Ford's Aerostar — were minivans too. True, the Astro, Safari, and Aerostar *were* a bit smaller than standard vans, but they missed the essence of what made Chrysler's minivan popular: its low, compact size and carlike feel, which Chrysler was able to achieve because the minivan was actually built on a car platform.

Whether due to hubris or excessive caution, Ford delayed for nearly a decade before introducing its Mercury Villager minivan, and GM watched Chrysler grab minivan buyers for half a decade before bringing out its Lumina APV minivan. Japanese automakers were also slow off the mark, though they felt hindered by the belief, which later proved erroneous, that their imports would be subject to a 25 percent tariff. (To the annoyance of the Big Three, the tariff on trucks was interpreted as applying only to those that primarily carried cargo. Vehicles that primarily carried passengers were exempt.)

For five delicious years, Chrysler had the burgeoning minivan market mostly to itself. In the minivan's very first quarter, Chrysler earned $705.8 million, more profit than the automaker had posted the entire previous year and the best results in the company's sixty-year history. By year's end Chrysler racked up nearly $2.4 billion in profit.

Rarely had a company so close to bankruptcy sprung back to health with the vigor of Chrysler. The publication of *Iacocca: An Autobiography* in the fall of 1984 further pushed Chrysler into the public eye. The book was a best-seller for thirty-seven consecutive weeks, all but one week at number one. In less than a year, bookstores sold more than two million hardback copies, which put *Iacocca* in a league with *Gone with the Wind, The Power of Positive Thinking,* and *Jonathan Livingston Seagull.* Overnight Iacocca began receiving as many as five hundred letters a day at Chrysler headquarters and an average of a thousand invitations every month to speak. Readers of *Iacocca* were fascinated by the legend of the immigrant's son who fought his way to the top of the corporate world, was booted out by black-hearted Henry Ford, and then, by dint of his courage and talent, managed to claw his way back. Though tremendously biased, *Iacocca*'s peek at what went on inside the executive suite was quite rare and, therefore, compelling. He preached the work ethic, patriotism, and devotion to family. Professing to be an admirer of the Japanese, he nonetheless attacked U.S. trade policy and prescribed get-tough tactics with Japan. Iacocca cleverly cemented his credibility as a storyteller and salesman by finally admitting what was never said during Congressional hearings but what all too many buyers knew: that Chrysler had built millions of awful cars.

Quite naturally, Chrysler gained visibility, attention, and serious consideration from readers of *Iacocca* who reasoned that so frank and honest a corporate leader would never be associated with any but worthy cars and trucks. Iacocca's story attracted the mass of everyday Americans, people who normally didn't read books. No business executive ever enjoyed a similar flood of hero worship. The phenomenon proved useful for the company. A gynecologist from Eufaula, Alabama, Dr. William Pappas, wrote Iacocca in 1985 to say that his next car was going to be a Chrysler. Philip E. Burr of Temple, Texas, wrote Iacocca to report that he and his wife had bought a Chrysler after reading his book. Similar letters poured into headquarters daily.

The automaker also received plenty of angry letters from customers who felt betrayed by their Chrysler vehicles. In fact, the quality of Chrysler's model line since Iacocca's arrival was no better than average—and often well below average—according to the number of defects counted by customers and reported in surveys by the J.D. Power automotive consultants. The quality of Chrysler's vehicles was certainly not approaching that of the Japanese models.

The original K cars, the 1981 Dodge Aries and Plymouth Reliant, were recalled three times during the first year of production to correct brake-light failures, sticky throttles, and hydraulic systems prone to water intrusion. Over the next three years the Aries and Reliants were recalled five more times to correct brake and other defects. Every automaker faced recalls, and Chrysler had its share. During the months when bankruptcy loomed Chrysler engineers and suppliers were under intense pressure to meet production deadlines while trying to comply with new federal regulations. The K cars were actually much improved over the Chryslers of the 1970s but of lower quality than the competition's.

Beyond the issue of defective parts and materials, the K cars exemplified Chrysler's reliance on outmoded automotive technology. The K car's suspension, for example, consisted of a solid rear axle to both wheels instead of the more advanced system, featured on some competitive models, of independent axles for each wheel. Thus, each time the Reliant's left wheel hit any uneven section of pavement, the vibration was transferred to the right wheel and to the rest of the car more strongly than would have been the case with a vibration-dampening, independent system. Buyers didn't have to understand suspension technology to discern immediately the handling difference in a car employing the more advanced system. Chrysler's basic four-cylinder engines, based on a twenty-year-old technology, were solid but old-fashioned. Four-cylinder engines were relatively fuel efficient because they were small, but they also generated relatively little power and lots of vibration. Thus, Chrysler cars felt rather shaky and anemic.

Nevertheless, the K car was utilitarian and relatively inexpensive; so it proved popular. The K platform that saved Chrysler from doom became the plow horse of Chrysler's model line. By stretching, adapting, tweaking the K platform, Chrysler was able to offer a dozen different models, including a sports sedan, a couple of station wagons, the minivan, an executive limousine, and others. Each model received a jazzy name: LeBaron, New Yorker, Laser. Iacocca realized that some buyers still wanted convertibles. Automakers, worried about safety, all had discontinued them. Voilà—he chopped the roof off the Reliant, added some chrome doo-dads, and the Dodge 600 convertible was born. The basic 1985 Reliant sold for about $7,000, the limousine for nearly $23,000.

Chrysler's model line represented Detroit ingenuity and chicanery at its finest. Iacocca took a basic, no-frills Reliant, added a flashy chrome grille, lined the roof with vinyl, stuck an ornament on the hood, swathed the interior with "Corinthian leather," and created a luxo-barge. With fake wire wheels, pinstripes, and a muffler that permitted more engine nose, the same car was a sporty model. Iacocca knew well that most American buyers didn't want sophisticated engineering, they wanted style. He understood buyers' fantasies, and he knew how to fulfill them automotively.

The research-and-development expense of bringing along a successor to the K platform was at least $1 billion, and possibly more if a new engine and transmission were included. The longer the expense could be delayed the better. If Lee Iacocca had learned anything at Ford, it was to squeeze the most value from existing technology. No sense in designing a new shock absorber if the old one did the job. Mustang buyers didn't see their car as a gussied up Ford Falcon with sexy new sheet metal, redesigned upholstery, and a fancy new dashboard; but that's what it was. Ford didn't need to design an expensive new chassis and power train when customers didn't detect the difference. From a financial standpoint, the expense of new mechanical componentry would have greatly diminished—and might have erased—Mustang's stunning profitability.

Hal Sperlich too had been trained at Ford and had learned, as Iacocca had, that cars had to make money, not just win admirers. But Sperlich was astute enough to recognize that the K car lagged far behind competitive Japanese models in technology. The K car had been built for one reason: to comply with tough fuel-efficiency standards. Front-wheel-drive technology was new for Chrysler engineers, and the K car had been a worthy first effort; but it lacked refinement.

True, sales of Chrysler family sedans based on the K platform were respectable enough in 1984. Demographically, however, Chrysler's customers were the oldest, least affluent, and least savvy about new technology of all domestic car buyers. Younger, more affluent buyers—precisely the buyers who were migrating to Japanese brands—were willing and able to pay higher prices. If Chrysler hoped to attract this group, it needed a better platform for a competitive family sedan, a project that required three to five years of engineering work. If Chrysler wanted to neutralize Honda and Toyota, Sperlich knew the project had to start without delay.

Iacocca didn't grasp the magnitude of the technology gap. His idea for injecting flair into Chrysler's model lineup relied on superficial rather than substantive changes. In 1984 he proposed an alliance with an old friend, Alejandro de Tomaso, who was running the Italian automaker Maserati. (In the early 1970s, Tomaso and Iacocca had created the Pantera, a car powered by a Ford engine, which had flopped because of safety concerns and limited interest from buyers.) Iacocca thought a sexy Italian two-seater, reminiscent of the Mercedes roadsters, might help overcome Chrysler's blue-collar image and attract better-heeled customers to showrooms where they might consider the more pedestrian K cars.

The venerable Maserati name connoted speed and luxury, so Iacocca proposed a hybrid comprised of Chrysler's K car platform with a body built in Italy by Maserati. It was a tried-and-true tactic from Detroit's bag of flim-flams. Buyers would *think* they were buying a Maserati, one of the authentic hand-built automobiles, normally priced at $100,000. Chrysler's Maserati would sell for only $30,000 or so. Iacocca didn't intend to sell more than 5,000 or 10,000 annually. The point wasn't to generate immense profits but to lend an air of style and panache to Chrysler's image.

For ordinary customers—the working families, young couples, and students whose income was no more than $20,000 or $30,000 a year—Chrysler needed an economical subcompact. The Dodge Omni and Plymouth Horizons, like the K cars, had served Chrysler admirably, but they couldn't hold their own, in quality or value, against the new Toyota Corollas or Honda Civics. Each year since 1981, the "voluntary" restraints were relaxed a bit more, and the Reagan administration had decided the time was drawing near when they should be abolished. Iacocca railed that U.S. trade policy for Japan wasn't tough enough, but he also believed that, given union wage scales and corporate overhead, Chrysler couldn't make a profit building a competitive subcompact in this country. Iacocca decided to pursue the same strategy as General Motors and import small cars from Japan.

Chrysler owned 35 percent of Mitsubishi, which it had bought in the 1970s, and had been importing Japanese cars for a decade. Though Mitsubishi's executives felt puzzled and betrayed by Iacocca's speeches calling for a tough trade policy against their country, they were nevertheless eager to export more cars under the Chrysler name. Mitsubishi's own retail dealer network in the United States was embryonic, and the Mitsubishi brand wasn't catching on quickly. To some Mitsubishi executives, Iacocca's behavior was proof of Detroit's confused approach to worldwide competition. If Iacocca was against imports, how did he square that opinion with Chrysler's own vigorous import practices?

Iacocca was asked the question over and over by reporters. His stock answer was: "I didn't make the rules, and I don't like the rules. But if that's what the rules are, then I'm going to play them to my advantage." Chrysler had attempted, on antitrust grounds, to torpedo GM's joint venture with Toyota to build cars in Fremont, California. But the Justice Department under Ronald Reagan was growing lenient in its interpretation of antitrust law, in accord with those who argued that concentration of economic clout no longer was a danger for American competitiveness. The government gave General Motors and Toyota a green light, prompting Iacocca to instantly pursue a similar venture with Mitsubishi. (In April of 1985, the two automakers announced they would build small cars, no later than 1988, at a location in the Midwest; a short time later they announced the formation of Diamond Star Motors in Bloomington, Illinois.)

Cars were but one portion of Chrysler's dilemma in the U.S. automotive market. Chrysler's full-size pickup truck, the Dodge Ram, was long overdue for a thorough mechanical redesign, as well as an interior and exterior styling makeover. Full-size pickups were an idiosyncratically American vehicle, the choice of the nation's plumbers, carpenters, gardeners and electricians. These buyers, overwhelmingly male, had overwhelmingly macho taste. Successful pickups had to be tough, powerful, and technically competent. They had to look tough as well as be tough. Japanese and European automakers never even attempted to build full-size pickups because no one except Americans would buy them and American buyers were loyal to

Detroit's brands. Without a competent full-size pickup truck platform Chrysler lacked the foundation for building a worthy sport-utility vehicle, like Jeep's Cherokee or Chevy's Blazer. The trucklike sport-utility was growing very popular among American buyers as a bigger, stronger alternative to the station wagon. "Sport-utes," as they were called, were really nothing more than a pickup truck with extra seats and a roof covering the rear.

By the early 1980s, Chrysler's Dodge Ram pickup had grown so outmoded compared to its domestic competition that engineers had nicknamed it "Festus," after the grizzled character from the TV series *Gunsmoke.* With its financial headaches Chrysler could never seem to scrape up enough capital to develop a new pickup; so Ford and General Motors stole the diminishing pool of Chrysler truck buyers and dominated this sizable and highly profitable segment. Sperlich and the engineers working for him desperately wanted to develop a new Ram pickup that could vie with the Ford F series and General Motors C/K series pickups. By 1985, with Chrysler's debt paid off and strong sales of the K cars and minivans, finances finally were strong enough to undertake a new midsize car platform and a new pickup truck.

Iacocca, however, had other ideas. Sales and profits were strong again, but if history was any guide, that prosperity wouldn't last for more than a few years. During this period GM and Ford were spending surplus cash as fast as possible to diversify their companies. The theory was simple: automaking provided cash flow during periods of economic growth; nonautomotive businesses would provide a steadier, noncyclical source of cash during recessions. In that way, the automakers' overall financial picture would be steadier and more predictable.

Wall Street liked predictability and steady growth. Hence, a diversified company's stock was valued more than that of an automaker. Typically Chrysler's share price was three times the annual per share earnings, while Ford's and GM's sold for four and five times earnings. The stocks of some diversified industrial companies were selling for fifteen and twenty times earnings. Those numbers made Iacocca's mouth water. His options were tied to the valuation of Chrysler stock.

Boosting the multiple from three to fifteen was a way to turn a nice nest egg into a small fortune.

However, if Chrysler wished to spend time and money on companies like Hughes Aircraft, Hertz, or First Nationwide, it would have to postpone development of new cars and trucks—of that much Hal Sperlich was certain. Chrysler was flush, but, in Sperlich's view, it didn't have the money for both. Iacocca was showing more and more interest in nonautomotive businesses and delaying more and more of Sperlich's requests for development funds.

During the summer of 1985, Chrysler teamed up with Allied-Signal, a diversified conglomerate, in a joint bid of $3 billion for Hughes Aircraft. Ford bid too; but it was GM's chairman Roger Smith, with far more money at his command than Iacocca, who won the auction with an offer of cash and stock worth over $5.2 billion. Hughes hadn't built aircraft in years, but it employed thousands of the nation's top engineers. Roger Smith was eager to acquire Hughes's vast high-tech resources and personnel in defense electronics, telecommunications, and specialties like advanced materials research. Smith believed that space-age scientific capabilities were going to be increasingly valuable to automakers as cars and highways became more sophisticated.

A month after losing the Hughes auction, Chrysler agreed to pay $637 million for Gulfstream Aerospace. Iacocca explained that buying Gulfstream was a way to diversify into "high-tech industries like aerospace and electronics." In fact, Gulfstream was a leading builder of corporate jets, whose owner, Allen E. Paulson, was a pal of Iacocca's. The acquisition didn't really fit Chrysler's diversification goals, except by the most liberal interpretation of the words *aerospace* and *high-tech.* But Iacocca's associates, who remembered how furious he had been when U.S. Treasury officials insisted he give up flying corporate jets during the bailout, snickered among themselves for they knew that as long as Chrysler owned Gulfstream, Lee Iacocca wouldn't be flying in anything but the newest, fastest, most lavishly appointed jet that Gulfstream could build.

Iacocca's sudden relish for the financial acrobatics of nonautomotive acquisitions discouraged Hal Sperlich. He recalled bitterly his endless battles with the finance men at Ford Motor Company, where he had fought to convince them to spend more on cars. Sperlich loved cars, loved designing, sketching his ideas, watching them evolve from paper to clay to metal. Ford's finance men had been conservative; they tried to prevent Sperlich from spending too much money on new designs and technology, preferring to improve the company's financial position by selling yesterday's cars for as long as possible.

Sperlich believed that Henry Ford too often favored the finance men, to the detriment of the company. He argued constantly with his boss for more advanced technology and snazzier designs. Each time Henry Ford said no, Sperlich tended to argue longer, louder, and in less compromising terms than anyone dared argue against Henry Ford, which undoubtedly contributed to Sperlich's departure.

Now Sperlich found himself once more fighting for development money, this time with Lee Iacocca. The executives would gather in the big fifth-floor conference room at the Highland Park headquarters — Iacocca, Bidwell, Greenwald, Miller, and Sperlich. He would present his case for a new engine or for accelerated development of a new car or truck platform. Over and over Iacocca turned him down. The cars needed a new "skin" or a new dashboard, Iacocca said, that was all. As the executive directly in charge of automotive operations, Sperlich felt undercut by Iacocca's skepticism. He couldn't tell whether or not the others agreed with him, for it was his impression that the other Chrysler executives weren't as demonstrative and didn't speak up about issues they believed that Iacocca had already decided. The others watched Sperlich and Iacocca debate but said little.

Sperlich prided himself on avoiding politics, on having the guts to speak up when an issue of principle was at stake rather than say the "right" thing. His talent as a car man had carried him far, but he had never really submitted to the deeper truth that one can't keep fighting the boss, especially in front of others, and survive for long in Detroit.

As Iacocca's corporate vision grew beyond the boundaries of the U.S. auto industry, he decided that Chrysler needed a charter and corporate structure more suitable to its new lines of business. This formality, he believed, might cause investors to value Chrysler's stock more highly, the ultimate grade on the company's report card. Instead of being organized primarily as an "automaker," Iacocca decided in late 1985 that Chrysler should be reorganized legally into a "holding company," with four main lines of business: Chrysler Motors, Chrysler Financial, Gulfstream, and Chrysler Technologies. This fourth line of business, Chrysler Technologies, remained an "empty box" on the organizational chart until Chrysler could find some companies to buy and place in it.

Likewise, Iacocca scrapped his four-member office of the chief executive and replaced it with a five-member corporate executive committee. Robert S. Miller, Jr.—known as Steve—was formally added to Iacocca's inner council of Bidwell, Greenwald, and Sperlich. Miller, a tall, Harvard-educated financial specialist who had been recruited from Ford, had played a central role with Greenwald in bailout negotiations with the government and lenders. Afterward, he was named chairman of Chrysler Financial, the fast-growing financial-services subsidiary. Miller's appointment to Iacocca's inner circle underscored the importance Chrysler's chairman placed on expanding financial services. The most recent acquisition had come a few months earlier when Chrysler bought a consumer finance company from Bank of America for $405 million. Chrysler Financial's assets through the mid-1980s doubled from $7.2 to $14.4 billion. What had been an in-house bank that raised funds and then lent them to dealers and retail buyers for the purchase of Chrysler cars and trucks was turning rapidly into a profitable financial-services supermarket.

As Iacocca distanced himself from Chrysler's businesses and appointed top executives to manage them, he began spending more time outside Detroit on non-Chrysler pursuits. In 1982, Iacocca was appointed head of the Statue of Liberty fund-raising and restoration committees by Secretary of the Interior James Watt. The Reagan administration wanted to prove that volunteers could carry out a big project that otherwise would require a large government public-works outlay. Iacocca, who had received a great deal of public exposure during the bailout hearings, seemed like a good choice. Members of the committees, which were based in New York, and some Chrysler managers did much of the work; but Iacocca made himself available to deliver high-profile speeches. A major portion of the $230 million bill for the restoration came from individual contributions. Corporate donations were sizable as well, averaging about $5 million each.

Just as impressive as the number of corporations willing to contribute were the sizable number that wanted nothing to do with the statue—precisely because of Lee Iacocca. One friend of Iacocca's who worked on the fund-raising committee noticed that as Iacocca became better known, he had become more outspokenly critical of government policy, more cocksure of his pronouncements. The nation's chief executives were a fairly tight-knit fraternity, many of whose members knew one another from serving on each other's boards or from groups like the Business Roundtable. Several of the chief executives solicited by the committee weren't entirely comfortable with Iacocca's tendency to grandstand and pontificate; some were friends or admirers of Henry Ford II and had disliked Iacocca's vengeful treatment of Ford in his book.

Iacocca squirmed over the possibility of rejection by people some might regard as his social betters. Hence, it wasn't easy for him to telephone certain fellow chief executives to solicit contributions. When there was a fund-raising meeting at the White House, his hands would shake. People working close to him were often surprised to see how insecure this commanding figure really was. Once, after a tough day of asking for money, he and a friend from the committee repaired to Iacocca's suite at the Waldorf Towers in New York for drinks. As Iacocca began to pour a bourbon, he suddenly blurted out: "These WASPs don't really understand us."

It was a recurring theme for Iacocca when he was in his cups or feeling thwarted: the WASP establishment, the Henry Fords of the world have everything handed to them. They don't understand rejection or humiliation the way Italians and other ethnic minorities do. Even within Detroit's automotive community, he was painfully aware that he belonged to the new money of Bloomfield Hills, not to the old money of Grosse Pointe, where the Fords and Dodges lived. Iacocca, so successful and so highly acclaimed for his accomplishments, was in many ways unsure of himself as he had been many years ago in Allentown when the other boys had ridiculed his Italian heritage.

Just as it became clear in February of 1986 that Iacocca's fund-raising had far surpassed expectations and would be celebrated at a smashing July Fourth spectacle in New York, he received a telegram from Secretary of the Interior Donald Hodel, informing him that he was fired as head of the Statue of Liberty–Ellis Island Centennial Commission. To say that Iacocca was shocked and outraged would be the epitome of British understatement. He called a press conference to blast Hodel and the Reagan administration; and he permitted friends such as Mario Cuomo and Frank Sinatra to lobby Reagan on his behalf. But it was all to no avail: this was a Washington-style beheading, a bit more genteel than the way it would have been handled in Detroit, but no less final.

For someone so shrewd, Iacocca was remarkably insensitive to the feathers he had been ruffling and to the true reasons for his firing. Officially, Hodel explained that the administration, while grateful for Iacocca's participation, wished to avoid a conflict of interest between the committee that raised money and the Centennial Commission that spent the money. The administration wanted Iacocca to see the fund-raising through to the end; someone else would lead the committee to restore Ellis Island. What Hodel failed to say was that Iacocca had infuriated powerful administration and Republican figures with his frequent attacks on Reagan's economic and trade policies.

Iacocca had alienated Don Regan, the Secretary of the Treasury and, afterward, Reagan's chief of staff, in the quarrel over the warrants for Chrysler stock. Regan and others couldn't believe Iacocca's cheek in asking the government to forego the warrants. The two had had rough words. Iacocca had tried to bully Regan, who hung up on him.

By that time, though, Iacocca had already been appointed as chairman of the fund-raising effort and there was no graceful way to get rid of him. Then *Iacocca* hit the best-seller list and his ego seemed to swell to new and epic proportions. There were few subjects on which Iacocca wasn't willing to pass judgment, particularly when it came to how badly he thought the American government was being run. Department of Interior officials muttered bitterly that he was running the fund-raising and restoration committees as if they were his personal staffs, firing off orders and paying scant attention to opinions that differed from his. When pollsters and campaign organizers in Washington began urging Iacocca to consider running for president (and he did little to discourage their urging), administration officials decided the time had come to trim Iacocca's horns. The moment the campaign goal was reached Iacocca was doomed. It simply became a matter of carrying out the execution.

The episode was reminiscent of Iacocca's rumble with Henry Ford. Iacocca had let power and ego blind him to exactly where he stood in the power structure. Just as it had been at Ford, everyone around him hopped to his command or, at least, was extremely deferential. No wonder he grew accustomed to expressing whatever was on his mind with little thought about the fallout. Iacocca had done nothing wrong except let hubris obscure the consequences of publicly criticizing a sitting president while serving at that president's pleasure.

THE GILDED AGE

FUNDRAISING FOR the Statue of Liberty and the celebrity generated by Lee Iacocca's autobiography consumed time and energy Chrysler's chairman otherwise might have devoted to the automobile business. Increasingly, Iacocca's mind turned to loftier concerns than cars and trucks. As politicians, bankers, television commentators, and others sought his advice, he accepted more invitations to deliver speeches. He thought more about government policy and explored ways to expand Chrysler's size and stock value. Instead of mundane worries about automotive market share and warranty costs, Iacocca's concerns dwelled primarily on the prosperity of the nation, the future of education, and the problems of cities. He started writing a second book in which he explained his ideas on everything from diplomacy toward the Soviet Union to parenting.

As a businessman, Iacocca enjoyed the privilege of pontificating about national issues without arousing criticism that he was serving personal political ambitions. To average Americans, Iacocca's nostrums reinforced his image as a public-spirited, can-do executive, someone who could surely slice through Congressional gridlock if given the chance. Periodic polls of registered voters indicated that Iacocca's name indeed commanded a stronger positive response than many well-known politicians. Democratic party political consultants, searching desperately for a new face to break the Republican hold on the White House, tried to enlist Iacocca in the spring of 1986 for a presidential bid. Among them was Greg Schneiders, who had worked for Jimmy Carter. Iacocca declined, but only half-heartedly, thus heightening the enthusiasm of Schneiders and others who were trying to recruit him. He told interviewers he didn't think he would be any good at politics and that he was too old to learn a new profession.

Privately, Iacocca was flattered that so many people thought he might make a good president. He quietly sent assistants to New Hampshire to assess his chances and would mention casually to interviewers he could be elected president if he wanted. But when he considered the effort required to master politics, the sheer drudgery of nonstop campaigning, the loss of privacy, Iacocca worried that running for office would be awfully unpleasant. Republican political consultants gently reminded Iacocca that politics, unlike business, can be tremendously confining. Voters have proven to be much more fickle and vindictive than shareholders. By late July of 1986, in the afterglow of the Statue of Liberty festivities, Iacocca issued a straightforward public request to Schneiders and other Democratic political consultants to cease fund-raising activities in his name. But intimate friends understood that if he had ever wondered about his fitness to stand on a national stage and pass judgment on economic and political issues, his flirtation with presidential politics erased any doubts.

Nevertheless, Iacocca's influence was most impressive within the realm he knew best, corporate America. He had only to look around him to witness the trend to mergers and acquisitions that was transforming medium-sized companies with so-so balance sheets into vast empires — and large companies like Chrysler into international conglomerates. Companies whose stock prices were depressed because they did not perform well were being gobbled up by companies strong enough to cover acquisition costs with bank borrowing or with high-risk debt. The financial and operating weakness of many venerable corporations, whose reputations were stronger than their intrinsic value, left them ripe for the picking. With potential profit to be made by acquiring companies at a discount, a new breed appeared: the corporate raider. Raiders were wealthy tycoons who, with the help of merger specialists from law firms and investment banks, threatened to buy companies and, against their will, break them up for the value of their assets, perhaps realizing a profit in the process. Sometimes raiders merely threatened acquisition, agreeing to retreat in return for millions of dollars in "greenmail."

The stock market crash of 1987 and the prosecutions of Ivan Boesky, the notorious stock trader, and Michael Milken, who perfected the use of the high-yield debt securities that came to be known as "junk bonds," cast a new light on the merger frenzy that had reigned for nearly a decade. Following the crash, Lee Iacocca joined the legions of corporate chief executives proposing rules to slow the pace of corporate mergers. Chief executives of large corporations who were accustomed to swallowing smaller companies with barely an afterthought were suddenly frightened to find their own enterprises in jeopardy. At Iacocca's urging, Chrysler's board approved a "poison pill" provision to its bylaws that made its acquisition unattractive. The provisions would be triggered in the event a raider tried to amass more than 10 percent of Chrysler stock. How, Iacocca wondered, could businessmen who knew little about the companies they were buying presume to meddle in great corporations like TWA, US Steel, and Goodyear? Citing the need for America's economy to compete more effectively against the economies of Germany and Japan, Iacocca disparaged "bad" mergers, which he defined as those motivated mainly by money, without a thought toward improving the acquired company. "If I take $3 billion in capital and go out and buy Burger King," he wrote in *Talking Straight*, "you've got a right to ask me what in hell I know about hamburgers and french fries."

Iacocca nevertheless felt as entitled to buy companies as the raiders and tycoons who were continually scouting for acquisitions. With Chrysler having recovered from its financial troubles, Iacocca realized he was in a position to use the company's equity and his own credibility to leverage a merger. For most of 1986, Iacocca's trusted lieutenant Bennett E. Bidwell was quietly shuttling back and forth to France, negotiating with Renault to buy American Motors. The price of Chrysler stock was at an all-time high, giving the automaker the ability to trade its shares or borrow against its shares to buy companies whose assets were selling relatively cheaply.

At a meeting with the Chrysler board's finance committee and Peter Cohen of Shearson at the Bloomfield Hills Country Club in early 1986, Iacocca reviewed a proposal to buy E.F. Hutton, the securities brokerage house that was lately weakened by allegations of check-kiting. Iacocca and Steve Miller were eager to expand Chrysler's budding financial services subsidiary. Miller was a finance specialist. To him, owning a retail securities brokerage seemed like a less risky way to improve Chrysler's returns than to inject more cash into the automotive business. When profits from automotive investment took their cyclical dive, profits from financial services would keep overall corporate earnings stable. Or so the theory went.

Chrysler directors, however, were skeptical. Hutton was for sale because it was in trouble. Chrysler executives knew Detroit and auto-making; more than one newcomer to Wall Street quickly found his pockets picked clean. Malcolm Stamper, the retired chairman of Boeing, told Shearson's Peter Cohen, "If E.F. Hutton is such a good deal, why don't you buy it?" (Shearson eventually did, just before American Express gobbled up Shearson.)

American Motors, the second acquisition possibility, seemed more reasonable to the Chrysler board's finance committee. Handicapped by its tiny size and hurt by the tepid consumer reaction to its quirky cars, American Motors for years had been operating on the margins of the U.S. auto industry. Among its legacies were the utilitarian Rambler, the weirdly spherical Pacer, and a presidential candidate, George Romney, who ran the automaker in the early 1960s before becoming governor of Michigan and a presidential candidate. When American Motors floundered in the late 1970s, Renault came to the rescue and wound up owning nearly half the automaker. Having failed as an exporter of cars from France, Renault hoped to distribute, through American Motors, French-designed, Canadian-built cars in the United States. Renault's Encore and Alliance, however, though light and fuel-efficient, were no match for the wave of superior cars from Japan. The Alliance and Encore, moreover, suffered from well-documented claims by consumer groups that they were poorly constructed and had poor reliability records.

From Iacocca's perspective, American Motors had only one attractive asset, Jeep—or more precisely, the Jeep name and franchise. Research showed that American buyers had very positive opinions of Jeep because they continued to associate the brand name with the military vehicle that had helped win World War II. A slavishly loyal population of owners year after year bought Jeep's plush four-wheel-drive Wagoneer and its smaller cousin, the Jeep Cherokee, which could be driven on rough terrain just like the military model. Young drivers liked the Jeep CJ7, though it had an unfortunate reputation for overturning and injuring or killing its occupants. As enthusiasm for four-wheel-drive vehicles gained momentum among car buyers during the 1980s, Iacocca calculated that whoever owned Jeep would have a big advantage over other makers of four-wheel-drives.

Best of all, Jeep buyers tended to be young, well educated, and affluent. Jeep Cherokees were commonly seen in garages in Greenwich and Beverly Hills next to BMWs and Mercedeses. Customers tended to buy them "loaded," that is, with many pricey options like leather upholstery, fancy paint jobs, and sophisticated stereo systems. Buyers of Chrysler's K cars were just the opposite: older, less educated, and inhabitants of blue-collar neighborhoods. K cars tended to be bought with few luxury options. A longtime axiom of the automobile business was that keeping customers was easier and less costly than winning new customers; Chrysler's customers were literally a dwindling breed; Jeep represented youth and the future.

By 1986, Renault found American Motors' losses unsustainable, and the French concern was in no mood to invest more. The French government, strapped for capital, was encouraging big national industries like Renault to seek private ownership. In its own evaluation of American Motors, Chrysler decided it wanted Jeep and perhaps the plants where Jeeps were built but nothing else. Renault, however, knew that its best hope for recouping some of its stake in American Motors was a package deal that included that automaker's less attractive assets as well. Bidwell and Miller, despite repeated attempts, weren't able to budge the French on that point.

Chrysler might have waited for financial pressures on American Motors to force Renault to cough up Jeep. But if American Motors were forced into bankruptcy, buying Jeep might be tricky and Jeep's image could lose its luster. Waiting would also have allowed a competitor, possibly Ford, to pick up American Motors. Renault executives were keeping the negotiations secret from their own U.S.-based executives, even from Joe Cappy, AMC's president. If talks were to become public, other bidders might crowd Chrysler out.

After negotiating for the better part of a year, the talks abruptly halted in November 1986 when Georges Besse, Renault's chairman, was murdered in Paris by political extremists. A few months later the talks resumed, Renault now led by Besse's successor, Raymond Levy, who was also opposed to selling Jeep separately. Rumors were swirling, but Levy maintained silence, insisting to Cappy and other directors of American Motors that AMC wasn't for sale.

As Chrysler executives realized the only way to get Jeep was to buy all of American Motors, it became apparent that some of them had deep misgivings. The Jeep franchise undoubtedly was a prize, but along with Jeep came the fixed costs of American Motors. Chrysler wasn't going to be able simply to fire 20,000 American Motors managers, engineers, and workers. Chrysler would have to accept billions of dollars in pension liabilities, legal liabilities, plants, offices, machines, real estate, and other expenses. And the lawsuits relating to Jeep rollovers were a nightmare in and of themselves. By merging with AMC, Chrysler would be returning to its condition in the 1970s, when high fixed costs threatened to overwhelm the company.

The most strident opponent to the merger was Hal Sperlich. For three years Sperlich had been nagging Iacocca to loosen Chrysler's purse strings to develop new vehicles, especially a new platform for a family sedan. The L/H platform was at an early stage of development, but Iacocca wasn't prepared to speed up the engineering because, in his view, there was still plenty of mileage in the K cars. Iacocca thought that the so-called new trends in styling were absurd. Ford's Taurus and Sable sedans, introduced in 1985, were the funniest things he had ever seen. He called the Taurus "a flying potato" and predicted it would flop. Sperlich knew better. Taurus's sales figures were impressive and the car, despite some early quality problems, was receiving good reviews. Its rounded styling made Chrysler's squared-off models seem all the more old-fashioned. In Sperlich's view, that Iacocca didn't recognize the Taurus as a serious force among family sedans helped explain how he could be more interested in buying American Motors and Jeep than in spending money to develop a car to compete with Taurus.

Bidwell was also against a merger with AMC. He too feared that Chrysler would be taking on hundreds of millions of dollars — perhaps billions of dollars — worth of potential liabilities stemming from lawsuits from accident victims in rollovers of the Jeep CJ models. Bidwell was also wary about AMC's new Premier sedan, to be built in Bramalea, Ontario. In his view the Premier was not a worthy entry in the family sedan market. By buying AMC, Chrysler was committed to producing the Premier and buying the French-built engines and transmissions from Renault.

Greenwald and Miller were on Iacocca's side. Lutz voted for the merger too but had reservations about the additional costs that AMC meant. Iacocca announced to the five executives that his vote was the one that counted. In March 1987, Chrysler announced it had reached an agreement in principle to buy the American Motors Company, including Renault's 46 percent interest, for $1.5 billion. Along with the Jeep brand, Chrysler also assumed control of a new assembly plant in Bramalea, Ontario, a network of 1,400 AMC-Jeep dealers, outmoded assembly plants in Toledo, Ohio, and Kenosha, Wisconsin, and nearly $800 million of American Motors debt. The final agreement also included a provision that forced Chrysler for a few years to continue building a version of AMC's soon-to-be introduced Premier with engines and transmissions from Renault.

Issuing $1.5 billion worth of Chrysler stock for American Motors and assuming AMC's financial obligations was clearly going to increase Chrysler's fixed costs drastically and push the introduction of new models further into the future. The merger at least raised the hopes of the workers at the American Motors plant in Kenosha, Wisconsin. Soon after the merger was disclosed, Chrysler announced it would continue to build cars there for up to five more years.

Meanwhile, after nearly four years of robust demand, the American auto market was showing signs of losing steam. The number of cars sold by Detroit automakers in 1986 was high but the figures were somewhat deceptive, as they were being sustained by discounts and rebates. Revenues were high but profits were eroding. The car market was like a balloon with a leak that automakers were attempting to keep inflated by pumping in more air. Chrysler earnings were down in the fourth quarter of 1986 and again in the first quarter of 1987, as the automaker spent $100 million more than it had budgeted for rebates.

The costliest discounts—and the potentially most damaging— were those the automakers were allowing to Avis, Hertz, National, Thrifty, and other rental-car companies. The rental-car companies normally bought hundreds of thousands of cars at a time, rented them for a year or so, and then replaced them. Rental-car companies also provided automakers with an important safety valve: the Big Three knew that when their inventories of unsold cars grew too large, they could shrink them by offering deep discounts to the rental companies, which were only too happy to accept a financial incentive to replace their fleets sooner than planned. As a further incentive, when retail sales were particularly slow, the Big Three sometimes guaranteed to buy back the low-mileage Dodge Dynastys, Ford Tempos, Chevy Luminas, and other models from rental companies after only three to six months. That way, Hertz, Avis, and the others were relieved of most of the financial risk inherent in buying and selling cars.

As rental-car companies became important volume buyers of cars, the Big Three bought stakes in nearly all of them to assure themselves ready outlets for excess inventory. (Because Avis, Hertz, and National were controlled by Ford and GM, Chrysler had to settle for the smaller players, Thrifty, Dollar, and Snappy.) In essence, the automakers were becoming rental-car operators themselves. Travelers who were pleasantly surprised to find nine-dollar-a-day rental deals with unlimited mileage were experiencing just one more manifestation of Detroit's deep discounting.

True, deeply discounted cars were a money-losing proposition for the automakers, but Detroit auto executives were betting that deep discounts averted the greater financial losses that would be caused by idling the assembly lines that produced the slow-selling models. In bygone years, automakers had been able to minimize losses during

weak periods by closing factories until demand picked up. But because of new union job-security provisions, automakers were increasingly obligated to pay workers generous unemployment benefits when
they were laid off. And large inventories of parts were already bought
and on hand.

As absurd as it seemed, selling cars at a loss figured to be less
costly than not building them at all. The UAW had achieved its longtime goal of transforming labor from a variable to a fixed cost, like the
cost of buying a building or a machine. No longer could auto executives blithely regard autoworkers as mere "tools," to be hired and
then fired in keeping with the demand for cars. The Big Three's labor
compact with the union was beginning to resemble that of Japan's
promise of "lifetime employment" for its workers.

It was a long time before Detroit automakers sensed the insidious
erosion deeply discounted cars were causing in the demand for new
vehicles. The automakers loved to remind financial analysts that selling their cars to rental fleets gave potential customers an opportunity
to test drive them. By the same token, renting cars at low rates tended
to cheapen their image. As rental companies cycled cars more quickly
through their fleets, automakers had to scramble to dispose of hundreds of thousands of used cars in almost-new condition—but priced
far below brand-new models. Chrysler and others shipped the cars
from the rental companies to mammoth auctions, reminiscent of live-
stock shows. Word soon spread, and savvy shoppers felt little incentive to buy a new Chrysler New Yorker for $16,000 when a New Yorker
with 5,000 miles on it, and with the warranty intact, could be had for
only $11,000.

Initially retail dealers were furious because "almost new" cars—
sold at factory outlets and tent sales—depressed the demand for new
cars. To mollify the dealers, the automakers encouraged *them* to sell
almost-new cars as well, and the dealers soon discovered that they
were as profitable as new cars. Incredibly Chrysler, Ford, and General
Motors had created a system in which they lost money twice, once
when they sold cars to the rental companies, and once again when
they sold the "almost-used" cars at auction to dealers. Bob Lutz
called almost-new cars "the monster we let out of the box." Since

sales to rental companies were being counted as new-car sales, revenues, of course, looked great. Moreover, deep discounting to daily
rental fleets helped delude auto executives into thinking that they
were fighting the intrusion of Japanese models, which were gaining in
market share.

The mushrooming sales to Hertz, National, Avis and the smaller
rental companies were costing the automakers untold billions of dollars while creating the impression that Detroit's vehicles — once symbols of power and prestige — were being marketed as commodities,
like hog bellies or soybeans.

Chrysler's clouded financial picture and the burden of the American Motors acquisition contrasted sharply with the announcement in
April of Iacocca's own compensation, which soared to an unprecedented level for an auto executive. In the early 1970s the average
U.S. manufacturing executive was earning forty to fifty times the pay
of a factory-floor laborer. By the late 1980s the best paid chief executives were earning 200 times what ordinary workers in their company
were paid. According to Chrysler's proxy statement, the automaker
paid Iacocca $23.6 million in 1986, consisting of $14 million in salary, bonus, and stock, and another $9.6 million worth of exercised
stock options. Iacocca's payout, more than 400 times what an average
Chrysler worker was earning, reflected both the terms of an employment contract reached with Chrysler's board a few years earlier and
the rising value of Chrysler stock, which inflated the value of his
options.

Aside from a ritual Bronx cheer from the union, the disclosure
of Iacocca's compensation engendered little controversy, given the
context in which it was awarded. Stupendous sums were being made
on Wall Street. Steve Ross of Warner took home annual pay packages
measured in the hundreds of millions and Michael Eisner of Disney
and Robert Goizueta of Coca-Cola were earning $50 million to $100
million each. And these men weren't credited with saving their companies, as Iacocca was. In the eyes of most Chrysler shareholders,
Iacocca was worth every penny. His annual compensation was one
more reason, if he needed one, why the American presidency was not
an attractive option.

BOB LUTZ'S STORY

ONE EVENING in the spring of 1990, Bob Lutz stepped out of his suite on the fifth floor of the K. T. Keller Building, walked down the plushly carpeted hall of offices occupied by senior executives of the Chrysler Corporation, and pressed the button for the elevator to the garage. Waiting in his parking space in the basement was his freshly washed and fueled Chrysler Imperial. Within two minutes Lutz was gunning the customized luxury sedan along Oakland Avenue and onto the entrance ramp to Interstate 75, the Chrysler Freeway, for the forty-mile trip across Detroit and Wayne County to his house outside of Ann Arbor. No one had an Imperial like Lutz's. He had heavily customized it—for example, he had removed every trace of chrome—as a poke at Iacocca's taste, which he regarded as execrable.

As usual, Lutz was driving fast. Driving fast was a habit, but it was a habit he neither worried about nor intended to break. He had always loved the thrill of speed and the sense of freedom it conferred. Nothing interested him more than mankind's quest for motion and the machines that moved people from one place to another. He had flown fighter jets in the Marine Corps. He enjoyed Alpine skiing, competing in car races, riding motorcycles, flying his helicopter. Pushing machines to the edge to see what they could do led, naturally, to the occasional mishap. Lutz had crashed his helicopter once and was lucky to walk away from the wreck. Over the years, Lutz had become adept at talking his way out of speeding citations, usually by emphasizing his clean driving record. His high visibility in automotive circles helped him a bit too. Traffic cops across Michigan knew his face, or at least his name, from automotive enthusiast magazines and were eager to discuss new engines, models under development, or what Chrysler was doing to improve the police cruisers bought by their departments. Impressed by Lutz's macho charm, cops tended to send him on his way with a friendly warning to slow down. His last citation had been issued by a Michigan state trooper on I-94. He and the trooper shot the breeze about cars and flying and it turned out the trooper was a former helicopter pilot who had served in Vietnam. Lutz got the ticket, but the trooper showed up at the hearing to recommend a suspended sentence, with six months of safe-driving "probation."

This evening Lutz was driving fast for a different reason. He was venting his emotions by flattening the Imperial's accelerator. For months he had been frustrated, and now he was truly upset. Lee Iacocca, chairman of Chrysler and Lutz's boss, was making life miserable for him and for the engineers and designers under his supervision, who were doing their damnedest to prevent the automaker's collapse. Lutz and his team were trying to replace mediocre cars, reiterations of the old K car that Iacocca had clung to for so long. The Aries and Reliant had been just fine ten years ago, in the midst of the energy crisis. Plain and reasonably reliable, they represented basic blue-collar transportation at an affordable price. But performing cosmetic surgery on the K cars and trying to sell them to affluent buyers as Dynastys, Spirits, Acclaims, Shadows, Sundances, and Imperials had been a grievous mistake. Those models were meant to lure consumers away from Japanese models, but they didn't stand a chance in hand-to-hand combat with the likes of the Honda Accord, the Toyota Camry, or the Lexus LS 400. Consumers saw through the charade, forcing Chrysler to discount the vehicles until they were cheap enough to attract "price-sensitive" buyers. In other words, the cars were becoming harder and harder to peddle, except at heavy discounts, and rebates were shrinking Chrysler's profit.

There was still time for Chrysler to save itself, Lutz was convinced, but only if the company made a clean break with car models that weren't selling. Chrysler had to bring out better-looking, better-performing machines. Lutz saw this as his mission. The squared-off hoods and deck lids had to go. So did the vinyl roofs and the chrome "jewelry" that Iacocca loved to slather over the exteriors. The interiors had to be redone, with new fabrics and with controls that were easier to reach and operate. The potential was there. Tom Gale, the head of Chrysler styling, had been hankering for years to break away from the stodgy 1970s body and interior designs that Iacocca loved. François Castaing, vice president of engineering, and others working with him had already developed more advanced mechanical systems and engines and were testing some early prototypes of a new family-car replacement for the K car, called the L/H.

In addition to preparing a new midsize car, Castaing's engineers were even further along with a new four-door sport-utility vehicle, a project that Chrysler had inherited in the American Motors acquisition of 1987. Previously sport-utility vehicles were mainly two-door affairs, AMC's Jeep Cherokee being a notable exception. Ford's sport-utility, the Explorer, had proven that a well-designed four-door sport-utility could be both popular and profitable.

Instead of doing what he could to help the new projects along, however, Iacocca was acting more and more threatened by Lutz's suggestions that Chrysler needed an entirely new philosophy of automaking. Iacocca didn't question the need for new cars, but he resented the implication that his taste and judgment were passé. Of late, Iacocca was snubbing Lutz in meetings and dismissing his ideas abruptly. Iacocca showed a particular dislike for Castaing. Iacocca refused to explain his exact reasons, but clearly the Frenchman annoyed him. Iacocca seemed very suspicious of Castaing and what his engineers were doing; he was holding more private meetings with Castaing's managers, sending auditors in to harass Castaing's people. Iacocca was also holding "skip-level" meetings with midlevel employees in order to gather unfiltered information about what was happening in Chrysler's engineering labs. Theoretically, Iacocca's skip-level meetings had been instituted to expedite communication between senior executives and middle managers, but Lutz and Castaing suspected the meetings were being held to search for evidence of betrayal. Iacocca, Lutz suspected, was hoping to turn up some indiscretion or other petty excuse to justify firing Castaing, and perhaps himself as well.

Iacocca's irritation with both Lutz and Castaing was manifest in his delay of their pet project, the Viper. Iacocca delayed approving the $70 million needed to put the Viper into production, saying it was a lot to spend without assurance of a financial return. The powerful two-seater, to be sold for $50,000, was meant to ignite interest in Chrysler's Dodge brand among enthusiasts and the automotive press. Viper wasn't for the mass market; it would be built by hand and probably in such limited numbers that each Dodge dealer might only receive one or two to sell. However, publicity about the Viper in car magazines would count heavily among shoppers. If they liked the design — and Lutz was sure they would — shoppers would then visit showrooms to see the flashy car. Then they might end up buying a minivan or, later, one of the new midsize cars. But the Viper was important as more than a builder of customer traffic: Lutz hoped it would raise the spirits of the designers and engineers who were discouraged by the homely, unpopular cars that they had been commanded to produce. If Iacocca fired Castaing and others who had developed the Viper, Lutz feared the enthusiasm and energy among engineers to bring out new models might dissipate, endangering the automaker's very survival.

Finally, in May of 1990, Iacocca relented and approved Viper. Money was tight, and the chance for a measurable financial return was poor, but he couldn't deny that everyone at Chrysler and in the industry seemed to be dying for Viper to be built. By auto industry standards $70 million wasn't a lot of money, and the potential in improved image was fantastic. Chrysler's chairman, if anyone, could understand the benefits of a project like Viper. After all, Iacocca championed the exact same concept at Ford in the 1960s when he hired Carroll Shelby to spruce up that company's image with the Ford Cobra. It was Shelby, François Castaing, and Tom Gale, Chrysler's design chief, who had come up with Viper's design, and it had been the surprise hit of the 1989 Detroit Auto Show. Lutz suspected the reason for Iacocca's delay was the fact that Lutz had strongly championed Viper, so that it was seen as his car, not Iacocca's. Iacocca's own initiative to create an image-building vehicle, the Maserati TC, had flopped miserably, wasting nearly $500 million of Chrysler's cash in the process.

Because Iacocca had worked so many years at Ford, where he and other executives historically viewed Chrysler as a company always teetering on the brink of ruin, he was predisposed to wonder if Chrysler ultimately could survive without partners. For more than a decade, he had been nurturing a grandiose vision of a worldwide global automaker—"Global Motors"—with plants, engineering labs, and design studios able to supply any automotive market. He had acquired American Motors in 1987, and a year later toyed with the notion of making a bid for General Motors, going so far as to discuss the idea with Ed Hennessey, the chairman of Allied-Signal. The idea of swallowing GM was ludicrous, but it demonstrated to what lengths Iacocca was willing to let his imagination run wild to solve the riddle of how Chrysler was to survive for the long term.

Iacocca's next idea was to sell Chrysler to Fiat in an exchange of stock. A combination of the two automakers would more than double the size of Chrysler's playing field and establish Iacocca as a force in Italian and European society. Through the first half of 1990 he held talks with Giovanni Agnelli, Fiat's chairman, on the possibility of merging the two automakers. The Italian automaker needed new technology, more efficiency, and a way to gain access to markets outside southern Europe. By the same token, Chrysler was the only U.S. automaker without a presence overseas. Iacocca thought Chrysler could adapt a version of Fiat's Tipo rather than spend the money on a badly needed new small car to replace Chrysler's lackluster Sundance and Shadow.

Agnelli, however, was wary of joining forces with Chrysler, and he was wary of Iacocca. Agnelli had been Henry Ford II's friend and had been dismayed by the trouble his friend had gone through with Iacocca. Nevertheless, Fiat had its own troubles; though large and wealthy, it was a backward organization that had enjoyed government protection from competition in Italy and had failed to build a presence in the United States. As the two men discussed the merger of their companies, they became deadlocked over, among other things, how much Chrysler was worth and whether Fiat would be able to add anything more than money to Chrysler.

During the negotiations with Fiat, Lutz was telling anyone who would listen that Chrysler needed neither Fiat nor Fiat's Tipo; it could make its own small car. He also asserted that, with the minivan, Chrysler could start building a presence in Europe, slowly and deliberately, on its own. Part of Lutz's motivation for opposing the merger undoubtedly was personal; he badly wanted to run an independent Chrysler after Iacocca retired. He had worked thirty years to get to the top of an automaker and was now tantalizingly close to his goal. In a combined Chrysler-Fiat empire, who knew how Lutz would fare? Each time Lutz articulated his opposition to the Fiat merger, he knew he was alienating Iacocca. But he wasn't the only one opposed to Chrysler's loss of independence; François Castaing and many of Chrysler's engineers wanted no part of a merger with Fiat. To them, the Italian automaker had little technological expertise that Chrysler didn't already possess.

However the Fiat talks turned out, they were certainly a means for Iacocca to show Lutz that the chances were slim to none that he would ever be running an independent Chrysler as chief executive. By June of 1990, after a parade of executives left Chrysler, Lutz became one of two senior executives who seemed to be the logical candidates to succeed Iacocca. But Lee was being cagey; he already was a year beyond the customary retirement age and was vacillating constantly about his timetable. Indeed, it was Iacocca's ambivalence that convinced some of the executives that they might as well leave for greener pastures, since promotions might be blocked for years until the chairman's seat became vacant.

Lutz couldn't figure out exactly why Iacocca was so dead-set against him. When he tried to pursue the subject with him, Iacocca was evasive. On several occasions Iacocca said cryptically, "I've checked at Ford and they've got a thick file on you." Lutz didn't know what he was talking about, and Iacocca refused to be specific. What could Iacocca possibly have learned about Lutz that he didn't already know when he recruited him from Ford five years earlier? Iacocca had made no promises about advancement at the time Lutz joined

Chrysler. Lutz, meanwhile, had hung on and proven himself when many other Chrysler executives had chosen to quit. "Petersen and Poling told me all about you," Iacocca said. "They were happy to get rid of you, and they told me we'd be making a big mistake to let you be chairman." (Don Petersen and Red Poling, former Ford chief executives, later acknowledged that they had discussed Lutz's suitability as Chrysler's chief executive informally with Iacocca but said they did not render opinions against his fitness as chairman. Both denied the existence of a file containing damaging information about Lutz.)

Lutz wasn't sure whether Iacocca was telling the truth about Petersen and Poling. He had had run-ins with them at Ford before quitting to join Chrysler, but he couldn't imagine that his former bosses would have torpedoed his chances to become Chrysler chairman. In any case, he couldn't quite imagine why Iacocca would have gone to Ford for an evaluation on him when he himself had been fired by Henry Ford. Even if Petersen and Poling had disparaged him, Lutz couldn't imagine what kind of "file" they might be keeping. Maybe the "file" was just a figure of speech, something Iacocca dreamed up as a rationale for opposing his promotion to chairman.

Lutz pushed hard on the gas and the Imperial's engine launched the vehicle forward. In less than forty minutes he pulled into the driveway of his house near Ann Arbor, which he and his wife, Heide, had built to resemble a Swiss chalet. He parked the Imperial in his enormous garage, jammed with cars and motorcycles of every vintage. A 1937 MG; a 1952 Citroën; two BMW motorcycles; an old Dodge pickup truck.

Since the age of three, Bob Lutz knew he wanted to make cars. His father, a senior executive of the venerable Crédit Suisse bank, and his uncles had always owned racy, mechanically interesting cars: Alfa-Romeos, Aston Martins, Talbots, whatever was fastest and most beautiful they brought to his home in Zurich. Because his father's job took him for a few years to New York and then back to Zurich and then back again, Lutz was reared partly as a Swiss and partly as an American. He spoke German and French but was most comfortable in English. The constant change of locale played havoc with his education. Every time he moved back to Switzerland he was put back a grade, and he didn't earn his high school diploma until the age of twenty-two. Still, he was bright and had cultivated tastes for wine, cigars, food, and clothing, as well as for mechanically sophisticated motorcycles and cars.

As a young man, Lutz had an appetite for adventure, which attracted him to the Marine Corps, where he qualified for flight training and learned to fly several types of jet fighters. Apparently the speed and thrill of flying jet planes only enhanced, rather than supplanted, his enjoyment for steering a well-tuned car fast through a curve. While he was stationed in Florida and later in Texas he raced an MG-TD and an MGA in his spare time. After five years of peacetime duty in the marines, Lutz enrolled at the University of California at Berkeley, in the school of business administration. Concentrating on marketing, he would earn an M.B.A. from UC-Berkeley in 1962.

Upon graduation, Lutz was eager to get started in the automobile business. The Ford Motor Company recruiter who visited the Berkeley campus was impressed with Lutz and offered him a job. When he told his father that he was going to work for Ford, his father asked him, "Why not General Motors? Why not number one?" Lutz answered that General Motors hadn't sent any recruiters to campus.

The elder Lutz was acquainted with Fred Donner, GM's chairman. Donner was one of his contacts in the world of high finance in New York. A telephone call to Donner and Bob Lutz was invited for an interview and then offered an $8,800-a-year job as a planner in GM's overseas division in New York. Within two years General Motors transferred Lutz to its Adam Opel subsidiary in Germany, then to GM France, and then back again to Opel, as executive vice president of sales and marketing.

Lutz had quickly proved to be a dynamic, well-organized, and aggressive executive, traits that marked him for the fast track at General Motors. Lutz's charisma attracted people to him. He stood six feet four inches tall, kept athletically trim, and was gifted with cinematic good looks. Fond of telling stories and jokes, he charmed his colleagues, often by irreverently mimicking their bosses. When work finished and a group found itself at a tavern or restaurant — wherever in Europe he happened to be—Lutz was frequently the center of the fun, ordering wine from just the right vineyard, shmoozing with the innkeeper in his language, and then entertaining fellow executives with tales of aerial and racetrack derring-do. If such a thing as executive star quality existed, Bob Lutz had it.

Not yet forty years old, Lutz was put in charge of the marketing of one million GM vehicles annually in Europe. The job was quite substantial for so young an executive, a sign that bigger things were in store. He had one big frustration, though: he thought his pay was not commensurate with the job he held. Under GM's compensation system, pay and bonuses depended partly on seniority; because Lutz had been with General Motors for only eight years, his annual salary for 1970 was $27,000, a sum he considered far less than what he deserved. When he complained, his superiors at GM sympathized and promised him a big raise as soon as possible. But when President Nixon created a government wage-and-price board in 1971 to discourage inflation, Lutz was told he would have to wait. Then GM was hit with a financially costly nationwide labor stoppage, as a result of which the automaker decided to withhold executive bonuses. The wait for a pay increase seemed endless, so in 1972, putting aside a promising career at General Motors, Lutz responded favorably to an overture from BMW, the German automaker.

Resigning from General Motors, the world's premiere business concern at the time, was not something a fast-track executive did lightly. A GM executive post in the 1970s carried with it status and lifetime financial security. But low pay was not the only thing that bothered Lutz about General Motors. The auto giant was drifting further and further from the greatness it had achieved under Alfred Sloan. It began to show signs of turning into an overly cautious and bureaucratic institution, the sort of place where an executive might spend a lifetime without accomplishing anything significant. Meetings, reports, and financial analyses were valued more highly than cars. Lutz was not abundantly patient and didn't relish spending the next thirty years waiting for a chance to make his mark.

BMW, on the other hand, was emerging in the 1970s as a maker of powerful, thoughtfully engineered, trendy sporting sedans. Though tiny compared to GM, and not well known in the United States, BMW's 2002 model was beginning to show up on American highways and on the shopping lists of sophisticated car buyers. The carmaker showed great potential, and it was willing to pay Lutz half a million German marks annually, roughly ten times what General Motors was paying him. To Lutz, money was crucial, not just for what it bought, but for what it signified about his worth.

Lutz didn't work long as sales director for BMW before attracting the attention of Walter Hayes, who was in charge of Ford Motor Company's motor sports program in Europe. Lutz supervised BMW motor sports, so he and Hayes often ran into each other at races. Though Lutz possessed European sensibilities and could speak German fluently, he mentioned to Hayes that not being a native was always going to be a liability for him at BMW. That fact, he said, made him restless. Hayes recognized Lutz's charisma, his knowledge of cars, and his familiarity with the car market of Central Europe, where Ford wished to increase its penetration. Hayes recommended Lutz to Bill Bourke, Ford's head of European operations, as an ideal general manager for Ford of Germany.

Lutz was recruited by Ford in 1974 and within two years was promoted—by then president Lee Iacocca—to manage Ford's European truck operations in Great Britain. From the outset Lutz was anything but the picture of buttoned-down conformism that was the norm at Ford. He and the woman who became his second wife, Heide, made a fascinating, unconventional, and dramatic couple. And the sight of them riding their BMW motorcycles together, clad head to toe in skintight leather, only enhanced his bold presence and the popularity he enjoyed among Ford executives.

After Bill Bourke was transferred to the United States, Lutz reported to the new chairman of Ford of Europe, Harold A. Poling. Poling, known by the nickname Red, was a soft-spoken, methodical, detailed-oriented executive who was highly regarded at Ford and a favorite of Phil Caldwell, the man Henry Ford II ultimately chose to replace Iacocca. In some ways, Poling's personality was the antithesis of Lutz's. Poling wasn't the colorful, voluble subject of journalistic profiles; he liked cars well enough, but he was more at home poring over reports, letting the numbers speak to him. Working for Red Poling was no day at the beach for Lutz and some other Ford executives; in meetings Poling could be harsh and demanding with subordinates; but an hour later at a social function, all was forgotten and Poling turned warm and gracious.

Poling followed very much in the tradition of Phil Caldwell, J. Edward Lundy, and the Ford "whiz kids" who had emerged from the automaker's highly regarded finance department. These executives were trained to rely on analysis rather than intuition. By the time they were promoted to senior positions, they had been schooled in sales, marketing, and product development; they were well-rounded general managers, but still fundamentally numbers-oriented. And they had little tolerance for executives who flew by the seat of their pants, whose ideas were flash and sizzle and lacked the figures to back them up.

In business judgment as in personality, Lutz was completely unlike Poling and the others. Lutz saw himself as more intuitive and was prone to rely on subjective judgment and less willing to decide an issue on the basis of a statistical abstraction. Iacocca had been the same way, a marketing executive with an uncanny sense of what people wanted, how much they were willing to pay for it, and how to sell it to them. A table of numbers told Bob Lutz little; his talent, he thought, was his ability to look at a vehicle prototype and understand instinctively whether the model could sell. To Lutz, executives like Poling were fundamentally "left-brained," men who analyzed and reanalyzed every question, sometimes missing opportunities by failing to see the forest. He, on the other hand was "right-brained," creative and focused on the bigger picture. Poling and Lutz did appreciate each other's talents, but not without a generous measure of mutual suspiciousness.

By 1979, Red Poling was headed back to the United States, and Ford was ready to promote Lutz to run Ford of Europe. In less than five years Lutz had risen from relative obscurity at BMW to one of Ford's most important executives. He was being hailed by many as a genius, a boy wonder. By all accounts, Lutz truly made his reputation in Europe as an automaker with the Ford Sierra, a futuristic looking sedan that set the course of Ford's styling for the next decade. Ford executives in the late 1970s were faced with the choice of a new design for the car that was to replace the Ford Cortina sedan. The Cortina, a top seller in Great Britain, was conservative, even stodgy, like most of Ford's car models of that era. In Germany, however, it was unpopular. Ford designers wanted to add more features to the Cortina, like reclining seats, and to remove chrome—as BMW had done—but they could never get the go-ahead from the Ford brass, which was chronically fearful of alienating its faithful customers.

By the late 1970s, Ford's management in Dearborn finally decided that it had to change its cars in order to compete directly with Audi, Mercedes, BMW, and other German brands that were regarded as superior. Ford's European strategy was to become more dominant in Germany while holding on to its market leadership in Great Britain. Lutz ordered designers to build clay models of a Cortina-sized vehicle, but more modern, and of a second, far more daring version with completely new design themes. The second clay model satisfied what Lutz believed was the public's "latent" appetite for the avant-garde; the sedan was rounded, aerodynamic, unlike anything that had been seen on European highways. Ford's chairman, Henry Ford II, and Don Petersen, who was head of international operations and soon to be Ford's president, flew to Ford headquarters in 1978 to review the choices.

With both models covered and Henry Ford and Petersen sitting among Ford of Europe executives in the audience at the design studio, Lutz ordered the drape removed from the less stylish model. "This is what conventional thinking says we should build," Lutz said. "This," he said, removing the drape on the second model, "is the wave of the future."

Henry Ford and Petersen were initially shocked. The lines of the second model were a radical departure from what customers were accustomed to seeing. *Fat* and *bulbous* were two adjectives that sprang immediately to the minds of those who looked at Lutz's choice, not quite able to relate it to any conventional design. Ford, the carmaker, and Henry Ford in particular, had a reputation for conservatism. Until 1980, Ford's European division had built dependable, unexciting cars for the masses, cars that exhibited minimal design flair and only a few innovations from previous models.

Ford had always been strong in the small-car segment of the European market. The automaker made decent profits, but they were based on high volume and careful cost containment. Since Ford's larger and more luxurious models weren't highly rated by consumers, they didn't command the prices (or profits) of competitors like BMW, Mercedes, and Audi, particularly in Germany and throughout the more affluent northern Europe. The Sierra was Ford's tactical thrust into this more affluent, higher-profit end of the market.

For a conservative organization like Ford, the Sierra design was truly radical, as was the decision to drop the well-known Cortina name. Many longtime Ford executives were reluctant to offend faithful customers with strange shapes that they wouldn't associate with Ford. Lutz argued that buyers would in time get used to a new shape. A certain number of buyers, of course, were bound to hate it, but others would fall in love with the unusual design and want it. What must be avoided, in Lutz's view, were designs so safe that they were doomed to be everyone's second choices. Overcoming great skepticism, Lutz and other Ford of Europe executives convinced Henry Ford, Phil Caldwell, Red Poling, and Don Petersen to take a chance with the more innovative style.

Sierra's success was mixed. Some customers loved the styling, but automotive reviewers who hated it singled out Lutz for blame. At the same time, General Motors launched a new Cavalier model in Great Britain that competed more effectively, grabbing some share of the market that had been dominated by Cortina. With the introduction of Sierra and attendant publicity, Bob Lutz secured his reputation in the world automotive industry as a "car guy," someone whose bias was toward the daring instead of the safe and sometimes prosaic shapes. He himself saw the Sierra as a victory of the car guys at Ford over the more staid, sober automotive executives whose priority was to minimize risk.

Many in the auto industry were starting to understand that categorizations of auto executives as pure "car guys" or "bean-counters" oversimplified the issue of who was qualified to decide which cars were built. As the industry and the automotive market became more complex, senior executives didn't fall neatly into one category or another. Henry Ford, Don Petersen, and other Ford executives realized the importance of beautiful, exciting, practical automobiles as well as anyone else and regarded themselves as bona fide car nuts. Petersen, for example, had worked as a Ford product planner for much of his early career.

Lutz, by the same token, believed his enthusiasm for automotive design didn't detract from his credentials or judgment as a business executive. He understood as well as any financial manager the importance of creating cars that earned money as well as rave reviews. Successful carmaking was a triad, Lutz explained, of finance, technology, and art, and it produced fine automobiles when the three disciplines were blended harmoniously. The artistic, passionate element was often squeezed out because bean-counters were typically conservative. So carmaking had become homogenized and boring. But, on the other hand, neither should cars be so idiosyncratic that they failed to attract enough buyers. Creating great cars required a sense of balance, which Bob Lutz — and many of his admirers — believed he possessed.

While Ford's U.S. operations were mired in the recession of the early 1980s, Ford's European business was flourishing, a condition that many at Ford credited in large measure to Bob Lutz's leadership. Among fellow executives, the knock on Lutz was his high public profile and his easygoing relationship with the press. He was a great interview, a master of the colorful, self-aggrandizing quote. The worry was that Lutz was more interested in getting publicity for himself than for Ford, that he basically was a careerist who put his own ambitions before what was good for his employer. Phil Caldwell, who became president during the messy ouster of Lee Iacocca, ordered Red Poling "to turn Lutz off" following a particularly annoying spate of articles touting him as a future president of Ford.

In 1982, Lutz was promoted to executive vice president in charge of international operations and named a director, jobs that brought him from Europe to the Dearborn headquarters. At the age of fifty, he appeared to all the world as a contender, like Red Poling and Don

Petersen, to one day run the Ford Motor Company. And like many talented, ambitious executives before him—notably Lee Iacocca—Lutz was building a following at Ford, the necessary power base. But for a top executive he was more individualistic than was generally thought healthy at Ford. For example, at a companywide gathering of top Ford executives in Boca Raton, Florida, in the fall of 1984, the other executives delivering speeches showed up to rehearse their speeches in casual but "correct" attire. Lutz, who had been swimming laps in the hotel pool, showed up to rehearse his speech in wet swimming trunks. The conservative Ford managers at the conference were shocked, whispering among themselves about Lutz's daring. The swimming trunks weren't a gaffe in the eyes of his admirers but proof of how Bob Lutz differed from typical auto executives.

In *The Reckoning*, David Halberstam wrote that Lutz's individuality inside Ford made Henry Ford and some senior executives uncomfortable, leading them to wonder whether he might be turning into another Iacocca or perhaps into a renegade "glamor boy" like John DeLorean. Iacocca's firing, four years earlier, was still a fresh wound at the company. Perhaps Lutz did not fully appreciate the sensitivity of the Fords or realize that the family was not fond of executives who seemed to forget to whom the company belonged.

Lutz's career also ran into an obstacle rather more mundane than his excess of individuality: shortly after his promotion to executive vice president in charge of international operations, the European subsidiary began slipping. The slide seemed to have as much to do with a weakening European economy as with management errors; but in any event, cutting costs and dealing with events in Europe suddenly turned into a grim, difficult task for Ed Blanch, the new chairman of Ford of Europe, and Jim Capolongo, Ford of Europe's president. Lutz, as executive in charge of international operations, believed that Blanch, whose background was in accounting, wasn't able to manage under conditions of adversity and recommended that he be relieved of duty. Ford senior executives agreed, and Blanch retired. Capolongo, whose background was in product planning, became Ford's top European executive.

But the performance of Ford of Europe didn't improve under

Capolongo, and Lutz remained frustrated. He tried to give Capolongo pep talks, spoke harshly to him, but nothing seemed to work. As Lutz saw it, Capolongo was overmatched for his job and had risen to a top executive post as a result of his strong presentation skills rather than by demonstrating top-level executive judgment.

Many of Capolongo's friends and colleagues at Ford of Europe saw the situation quite differently. Capolongo was doing as well as anyone could under extremely difficult market conditions. Instead of helping, Lutz was doing very little constructive to improve European operations, preferring to fob off blame on the executives in Europe and thus shield himself from the fallout. It looked as if perhaps Lutz saw that poor results in Europe might hurt his career and he was trying to distance himself from European management. From his ivory tower in Detroit, Lutz had nitpicked Blanch and forced him out. Now it was Capolongo's turn.

In June of 1984 Capolongo abruptly quit. He hadn't warned Lutz or shown any sign that he was nearing his breaking point. Through friends, however, Capolongo leaked word that he could no longer bear working for Bob Lutz. The news of Capolongo's resignation surprised and upset Ford's senior executives, including Don Petersen, who was about to take over from the retiring Phil Caldwell as Ford's chairman. They had listened understandingly when Lutz recommended Ed Blanch's retirement, but Capolongo's resignation was too much.

Particularly upsetting to Petersen were unsubstantiated reports that Lutz's demanding management style was the reason the well-liked Capolongo was leaving. Petersen had become the champion of employee involvement and consensus-oriented management and was campaigning hard within Ford against the tough, intimidating tactics employed by many Ford managers.

Petersen decided to find out the facts firsthand. He flew to London and asked to meet with Ford's top European executives. In a series of meetings at his suite at the Savoy Hotel, one after another of the executives confirmed what Petersen had heard: Bob Lutz had ridden roughshod over Capolongo. What Petersen didn't know was that the executives had met among themselves before the meetings to synchronize their version of events; they were unhappy about Lutz's role in the resignation of Jim Capolongo and wished to make sure that Petersen understood clearly who they thought was at fault.

Lutz had been telling Petersen that Blanch and Capolongo were incompetent and the company was better off without them. After interviewing everyone, Petersen still wasn't certain how much of the blame was actually Lutz's, but he knew that Ford of Europe's top managers were thoroughly demoralized. Rectifying the "human relations problem," as Petersen called it, was Lutz's job. So Petersen, after consulting with Caldwell and others, ordered Lutz to return to Europe for a second tour as chairman of Ford of Europe, a somewhat awkward sidetrack to his career. Ford executives appointed Alex Trotman, a talented executive who was running the Asia-Pacific operations, as president in Europe, Lutz's second in command.

Lutz was none too happy to be returning to Europe. Even though he would remain a director, it was clear that he was not going to be named president, at least for the moment. That job was going to Red Poling. Lutz's wife, Heide, was very unhappy too. A handsome, headstrong woman, she thought Ford executives were being unfair to her husband and was angry that less than two years after moving from Europe they were being sent back. She and Lutz decided to maintain their primary residence near Ann Arbor and commute from Great Britain as often as possible. But Heide Lutz took another step, at the time, unbeknownst to Lutz: she telephoned Lee Iacocca to complain to the Chrysler chairman about how badly the Ford executives were treating her husband.

Shortly thereafter, Lutz received a telephone call from one of Iacocca's emissaries. The two men hadn't known each other well at Ford, but Lutz's infatuation with cars and his reputation among designers were well known to Iacocca. The Chrysler chairman had heard the rumors that Lutz's star at Ford had dimmed. Heide Lutz's telephone call raised the possibility that he might be able to exploit Lutz's unhappiness to his advantage and, once more, zing Ford by stealing one of its top people.

The two auto executives decided to meet secretly when Lutz was in Detroit for a Ford board meeting. Lutz didn't have to explain much about his situation because Iacocca had already heard the gory details. (Certain Iacocca loyalists at Ford made sure that their former boss stayed current on all important gossip.) Iacocca commiserated with Lutz and told him that he could forget about ever resurrecting his career at Ford, that he would never be forgiven and it was time to start thinking about coming to Chrysler. After all, who would know about the unforgiving nature of politics at Ford if not Iacocca?

Lutz tended to agree with Iacocca. He was interested in Chrysler. A key to Lutz's decision about Chrysler was the future of Hal Sperlich. At the time, Sperlich was Chrysler's top car guy, occupying the job that Lutz thought he was qualified to hold. How might Lutz improve upon the position he held at Ford when Sperlich was about the same age and ahead of him in the pecking order? Not to worry, Iacocca told him. Sperlich, talented though he was, didn't have the ability to take Chrysler to the next level. Sperlich was becoming more and more difficult to work with. Besides, Iacocca said, Chrysler board members lacked confidence in Sperlich and thought him unqualified for consideration as a future chief executive or chief operating officer.

Iacocca told Lutz that he planned to retire in about three years, after splitting Chrysler into a holding company. "The chairman's job is Jerry's [Jerry Greenwald's] to lose," Iacocca told Lutz. "If you do well, you can be chairman of the automaking operations."

Lutz was conflicted. He was unhappy about commuting between Detroit and Great Britain for his job at Ford. Heide Lutz wasn't very happy either. At the same time, Lutz would be damned if he was going to be driven out of Ford under a cloud. For reasons of pride, then, Lutz decided to try to redeem himself by sticking with the job of turning Ford's European operations around. Indeed, over the next year, the team of Lutz and Trotman helped position Ford to take advantage of Europe's improving economy. In 1984, the year Capolongo departed, Ford's European automotive operations achieved its highest-ever market penetration while suffering its first loss in history. In 1985, by dint of cost-cutting and reorganization, Ford of Europe's annual profit reached $326 million, a welcome turnaround. Lutz was eager to return to the United States and his home.

But Lutz's desire for a top executive post at Ford's Dearborn, Michigan, headquarters presented difficulties for Petersen and Poling, who were running Ford as chairman and president. Poling knew that Lutz was a favorite of Petersen's; but Poling had been adamant: Lutz had to earn his way back to the United States. He'd certainly done that, yet Lutz held the rank of executive vice president and director; and there was very little need in Dearborn for an executive of that stature.

Poling flew to Europe and told Lutz he had three options: stay in Europe for a few more years until a job opened up in the United States; return to the job of supervising all international operations; or take over Ford truck operations. Lutz didn't want to stay in Europe, so the first choice was unacceptable to him. But Poling also warned him that if he chose to run international operations from headquarters it would be his last job at Ford and that he, Poling, would be on him every minute. This was Poling's way of telling Lutz not to pick door number two.

Pete Pestillo, Ford's personnel chief, reinforced Lutz's impression that Poling intended to make things difficult for him. "You're an individualist," Pestillo told Lutz, "a swashbuckler. Ford really isn't the place for you."

Lutz, with little enthusiasm, agreed to run Ford's truck operations, a less prestigious job, but one that at least brought him back to the Dearborn headquarters. The press accounts of Lutz's less than triumphal return characterized his new job as a demotion. Truck operations had always been an automotive backwater, though their prestige was growing daily with the popularity of minivans and sport-utility vehicles, which were developed from truck chassis.

Lutz was not enthusiastic about working under the thumb of Red Poling. Just a few months after his return, relations between the two men were noticeably strained. Poling had snagged the job that Lutz wanted. Poling continued to hold Lutz partly responsible for what he regarded as the Sierra's mediocre performance. Finally, as the result of what may have been a bizarre misunderstanding, Poling accused Lutz of tapping Poling's phone, an accusation that outraged Lutz. (Poling acknowledged he conducted a brief, casual conversation with a

Ford of Europe executive and subordinate of Lutz's that grew out of a social call between the two men's wives. Somehow, Lutz had found out about the conversation and told Poling he was annoyed that Poling had breached the chain of command by speaking directly to one of his subordinates. Lutz's words led Poling to the conclusion that the call must have been monitored. Lutz vehemently denied the charge, but Poling refused to withdraw it.)

Lutz's acceptance of Ford's truck job was a means to return to the United States and to buy some time. Shortly after his return, in early 1986, Lutz, certain that Red Poling was determined to block his career, resumed his flirtation with Iacocca. Petersen and other Ford executives thought Lutz had redeemed himself admirably in Europe. Under his and Trotman's command, morale was restored and a big chunk was taken out of costs. Even Poling showed signs of warming up a little. He informed Lutz that he was receiving a $520,000 bonus for his performance in 1985. Lutz was quite pleased until he learned, a day or so later, that Phil Benton, a senior Ford executive in charge of diversified operations, had been awarded a $590,000 bonus. Why did Benton deserve more than I? Lutz wondered. Feeling miffed, he raised the issue with Poling, who told him, mysteriously, that the amount of Benton's bonus was connected to "personnel planning." That could only mean one thing, Lutz figured: Benton had passed him on the promotion list. The episode provided one more reason to entertain Iacocca's importunings more seriously.

For Iacocca, grabbing Lutz was a two-fold opportunity. The first, naturally, was to take a jab at Ford. The second was the chance to reinforce Chrysler's capability to conceive, design, and engineer new models. Having Lutz around would inspire the organization. Chrysler's design and engineering capabilities — the heart and soul of creating cars and trucks — had been largely the responsibity of Sperlich, Iacocca's longtime sidekick. But Chrysler's cars and trucks now needed more pizzazz. Lutz just might be the catalyst Chrysler needed to improve its vehicles' looks and performance.

Besides, Iacocca had learned about the care and feeding of executives from Henry Ford. Like Ford, he believed that adding one more strong executive to a house already full of ambitious egos was almost always beneficial. Another ambitious ego helped prevent others from becoming too complacent or from presuming too much about their futures. It was a technique Henry Ford had used repeatedly against Iacocca when he was nursing his ambitions to be president of Ford.

Before offering Lutz a job, Iacocca secretly delegated Jerry Greenwald to interview people who had worked for Lutz to discover if there were any skeletons in his closet. Iacocca knew that Lutz was a controversial figure and he wanted to be sure about him. When Greenwald interviewed Capolongo, he was told, not surprisingly, that Lutz lacked integrity and fired subordinates for his own mistakes. Some confirmed Capolongo's assessment, telling Greenwald that Lutz was self-centered, ego-driven, and looked out only for himself. Others spoke reverentially of Lutz, swearing that few car guys walking the earth were as talented.

What Greenwald learned from his research suggested that Lutz was a fairly typical auto executive: hard-driving, talented, and self-centered, no different from the other men who had left Ford to run Chrysler. Greenwald, who expected to take over soon from Iacocca, could only presume, of course, that Lutz would try to climb as fast and as high as he could on the corporate ladder, caring little for who got in his way. Lutz was a talent who had to be managed carefully. For the time being, the two men forged a truce. "Until you give me reason to feel otherwise, I'm going to trust you," Greenwald told him.

On June 3, 1986, Lutz gave Petersen his resignation. The same day, Chrysler announced that Lutz had been hired as executive vice president in charge of international operations (virtually nonexistent at that time) and of truck operations. Lutz also was elected a director. His portfolio wasn't large, for Iacocca expected Lutz to wage war and win what he could from Sperlich and others. Significantly, Lutz was to report to Jerry Greenwald, not to Sperlich, who was in charge of automotive operations. To those who understood the subtle undercurrents of power in the executive suite, Lutz's reporting relationship was a very bad sign for Sperlich's future.

Lutz's first battles at Chrysler had resulted in swift victories. He knew how to draw people to him and to make them believe in his automotive visions. Within eighteen months, Sperlich resigned and

Lutz was appointed top car guy for an automaker that seemed headed, in the eyes of automotive purists, nowhere. Besides the minivan, none of Chrysler's models were selling.

Now the 1990s were here and Chrysler's future was about to change. The Viper, the L/H family sedans, the Grand Cherokee were poised on the threshold, and Lutz was determined to make them winners. If only Chrysler didn't run out of money first. If only Lee Iacocca would get out of his way.

JERRY GREENWALD'S STORY

IN RETROSPECT, it's not hard to see why in 1990 Jerry Greenwald abandoned his dream of succeeding Lee Iacocca as chairman of the Chrysler Corporation.

Greenwald had been quite content as president of Ford of Venezuela and his career at Ford looked promising when Iacocca approached him in 1979 about joining Chrysler. A native of St. Louis and an economics graduate of Princeton, Greenwald possessed solid credentials as a finance specialist. Chrysler needed expert thinking about budgets and cost-cutting to solve its financial emergency. Greenwald and wife, Glenda, weren't crazy about leaving Caracas. Jerry, nevertheless, had to admit that he was attracted by the opportunity for adventure: Chrysler was an ugly mess that would test his talents and energy. He accepted Lee's offer.

For two years Greenwald and a band of Chrysler's financial and legal experts shuttled from meeting to meeting, from Washington to New York to Detroit, to negotiate concessions from government officials, union leaders, bankers, and suppliers. From a business perspective, no one was closer to Iacocca during that period than Greenwald. By the time automotive sales picked up and the emergency subsided in 1983, Greenwald emerged as Chrysler's vice chairman, second in command, and the man in charge of day-to-day operations.

From a career viewpoint, Jerry's decision to leave Ford and join

Iacocca seemed to have been prescient. Auto executives measured success by the distance on the personnel chart between their box and that of the chairman. Greenwald, who had set his sights on one day making corporate vice president at Ford—the holy grail for fast-track Ford executives—was now second only to Iacocca and the logical candidate to run Chrysler when Iacocca left the picture. If Greenwald had a weakness in the auto industry, it was that he, unlike his boss, had little feel for the tastes of car buyers or for automotive engineering. He was strictly a business executive and a general manager. When Iacocca left, others would have to worry about "product."

In 1983, when Iacocca's wife Mary died after a long bout with diabetes, Greenwald deduced from conversations with his boss that Iacocca wanted to retire. The bailout and his wife's illness had drained him significantly. But soon after the funeral, Iacocca abruptly changed his mind. "There's nothing for me now but my job," he told Greenwald, who was disappointed but not devastated by Iacocca's change of heart. After all, Greenwald was forty-eight, eleven years younger than Iacocca. In six years, Iacocca would reach the customary corporate retirement age and Greenwald would be only fifty-four—plenty of time to enjoy a good long run as chief executive.

Four years later, in 1987, after the publication of *Iacocca* and after all the Statue of Liberty hoopla, Iacocca again began dropping hints about retirement. He had plenty of money, and he enjoyed relaxing at his villa in Tuscany, puttering among the olive trees. It was clear that Iacocca didn't enjoy working long days any more, for he delegated many top executive tasks to others, preferring to spend his time delivering high-profile speeches, making television appearances, and pondering the possibility of a run at the presidency. Iacocca told Jerry Greenwald that he was ready to turn the company over to him.

But again Iacocca changed his mind. The stock market crash and the American Motors acquisition raised all sorts of pressing issues for Chrysler. Greenwald was disappointed a second time. But now only two years remained until Iacocca turned sixty-five. Greenwald was earning several million dollars a year in salary, bonuses, and options and was vested with complete operating responsibility for the company. Life really wasn't too bad. He could afford to wait for Iacocca's retirement.

In his eight years serving as number two, Greenwald learned the art of staying in Iacocca's good graces: sit, listen intently to Iacocca's monologues, and nod your head slightly. Never confront him, never contradict him in front of others. Iacocca was absent enough that Greenwald, by default, was able to leave a significant imprint on operations. But when he was around, Iacocca—and only Iacocca—was in charge. His proclivity for unilateral decision-making and long, windy orations was obvious even before Chrysler's turnaround. After the repayment of the government guaranteed loans and following the publishing success of *Iacocca,* his soliloquies grew longer and the illusion of omnipotence grew more firmly entrenched. The American public hailed him as its shrewdest, toughest businessman, and for his own part, he had few doubts about that judgment. If he decided Chrysler needed a two-seat European-style luxury car, who was to say it didn't? If he believed Chrysler should invest its cash surpluses in nonautomotive businesses, who dared argue that the money should be saved for a rainy day?

Hal Sperlich, Chrysler's president, relished his self-appointed role as counterweight to Iacocca on decisions about which cars and trucks to build and how they should look. Sperlich was rare among executives in his willingness to confront Iacocca. He was contemptuous of those, like Greenwald, who placated the boss. But even Sperlich fell into line when Iacocca was adamant. Steve Sharf, executive vice president in charge of manufacturing operations, sometimes was able to persuade Iacocca to listen to him; but Sharf always made certain to credit Iacocca for any success. In meetings with Iacocca, Sperlich, Sharf, and other senior executives unfailingly showed deference to the great man. For they sincerely believed—as did Jerry Greenwald—that without his leadership during the bailout, there would be no Chrysler.

In 1989, Iacocca again started making noises about retirement. He dropped broad hints in newspaper interviews about "being tired," about wanting to turn the reins over to someone else. For the third time, Iacocca told Greenwald he was getting ready to go. A year earlier Iacocca had created an "office of the chairman" comprised of himself, Greenwald, Bidwell, and Miller. And although he had not formally designated a successor, he had mentioned to enough people that the job was "Jerry's to lose." But when Iacocca celebrated his sixty-fifth birthday in October and showed no signs of stepping aside, Greenwald was at last convinced that Lee intended to run Chrysler indefinitely. The board of directors, several of whom Lee had nominated and dominated as its chairman, either couldn't or wouldn't force the issue.

Thus, when Greenwald's telephone rang in early January 1990 with a tempting offer, Jerry was ready to listen. Officials of the unions representing the mechanics, pilots, and others workers at United Air Lines wanted to buy the airline. The changing economics of air travel were hurting traditional carriers like United, whose managements wanted to reduce costs in part by cutting jobs and reducing compensation expense. The unions, in part to save themselves, believed an employee stock-ownership plan might reduce the bickering with management while giving United's workers a financial stake in its future. But an employee buyout required an enormous amount of borrowed cash, and the unions were hoping that finance expert Jerry Greenwald would be able to raise it. From his experience during the Chrysler bailout, Greenwald knew how to extract equitable compromises from the competing interests in a deal—the bankers, unions, suppliers, and government regulators. Ultimately, the hook of the union offer was that if the buyout succeeded, Greenwald would become United's chief executive.

The chance to run a major corporation as chief executive was Greenwald's dream. But United's buyout deal was chancy. Greenwald wanted to know why some airline executive hadn't jumped at this opportunity. The unions replied that they had tried to recruit Bob Crandall, the highly regarded chief executive of American Airlines, but Crandall had balked. And what would happen if Greenwald quit Chrysler and the employee buyout failed? His options to buy 215,625 shares of Chrysler stock were worth very little now, but that didn't mean they would always be worthless. Were the unions willing to reimburse him for the Chrysler options he would forgo if he left his job? For if the buyout failed, Greenwald was left without a job. To mitigate the risk of failure and to compensate Greenwald for the stock options he would have to leave behind at Chrysler, the unions were willing to meet an extraordinary compensation request: a $5 million signing bonus and a guarantee of $1 million annual salary, plus lots of stock options in the new corporation. And in the event the deal fell apart, the unions guaranteed Greenwald another $4 million, in escrow.

In late April of 1990, Greenwald telephoned Iacocca at his Italian villa. We've got to talk, he told Chrysler's chairman, I've decided to leave.

Greenwald's telephone call was unexpected and unsettling news for Iacocca, who relied heavily on Greenwald to run Chrysler so that he could relax as much as he cared to in Palm Springs or Aspen or Siena. In previous weeks, two other senior executives had announced their departures. One, Mike Hammes, left his job in charge of international business for a senior executive post at Black and Decker; and Iacocca had forced Fred Zuckerman, Chrysler's treasurer, to resign following differences on financing policy. Greenwald's resignation on top of these two might be perceived by the outside world—and investors—as a stampede.

Iacocca asked Greenwald to fly to Italy immediately. The next day, amid the grape vines and olive trees, the two men walked, talked, and played bocce. They recalled the bittersweet memories of the previous decade.

—*How could you walk away from the chance to lead Chrysler?* Iacocca wanted to know. It seemed to Greenwald that Iacocca thought he was crazy to consider leaving.

—*This is a huge challenge, Lee. United is in deep trouble unless it can be recapitalized and its costs reduced. They need someone who knows how, and there aren't that many people around with that kind of experience.*

Iacocca wanted to know if his own failure to announce a firm retirement date was the hangup. Would Greenwald reconsider if Lee were willing to announce a date immediately?

—Lee, I don't want to be the reason that you decide to retire. I need challenges. United will be a great one.

—Running Chrysler isn't enough of a challenge for you?

The tenor of the conversation had changed. Iacocca suddenly was no longer interested in declaring that he planned to retire and announce a timetable for management succession. Had he done so, Greenwald might have reconsidered. But after broaching the subject once, Iacocca didn't bring it up again. Perhaps he had been testing to see if Jerry was bluffing.

—You've made up your mind, Jerry, haven't you? Iacocca said.

Greenwald confirmed that he had. Because the financial press had detected rumors of Greenwald's talks with United—despite his denial—he felt he had no choice but to announce the decision immediately. Greenwald, Iacocca, and Iacocca's fiancée, Darrien Earle, left for the airport, where Chrysler's jet was waiting to return them to Detroit.

Greenwald's departure, which had initially struck Iacocca as catastrophic, was taking on a new meaning. Iacocca's plans for leaving, however vague, were suddenly, and unexpectedly, upset. With Greenwald's departure, Iacocca must have realized that, though he had lost a key executive, he had gained some leverage vis-à-vis his board. The directors had been encouraging him gently to announce a succession plan. The union, Chrysler dealers, and rank-and-file salaried workers were all growing anxious: What would happen if Iacocca left? Who would lead? With Greenwald gone, Chrysler needed Iacocca now more than ever. This was an ideal time to suggest a readjustment of his employment contract on more favorable terms. The directors, if they wanted him, had little choice but to give him more money and more flexibility to stay. The board couldn't blame Iacocca for the lack of alternatives. It wasn't his fault that Jerry had decided to go running after some quick cash.

A month after Greenwald's departure, Iacocca met with his directors to discuss what to do about succession. Following that June 7 meeting, he announced somewhat cryptically that he was staying "beyond the end of 1991," when his employment contract expired. How long? Under what terms? Chrysler spokesmen would only say that the issues hadn't been decided. Meantime, Greenwald's duties were divided among Miller, Lutz, and Bidwell, the last of whom planned to retire in January.

With no successor in sight, Iacocca knew he had the board over a barrel. The moment was perfect for squeezing extra pay and benefits in return for agreeing to stay on as chief executive, which is what he wanted in any event. On July 12 the board agreed to Iacocca's terms: in addition to his normal compensation, which had peaked at $23 million in 1986 but amounted to several million dollars annually, depending on the number of stock options exercised, he was entitled to *another* 62,500 shares of Chrysler stock for every quarter he stayed on the job beyond the end of 1991. Starting in 1992, he was entitled to an amount equal to the appreciation on 187,500 shares, valued at $15.50 a share, and any price the stock reached until five years after his retirement.

In other words, if the stock reached a price of twenty dollars in 1992, Iacocca would receive stock worth $5 million, in addition to his normal pay. If the stock reached, say, an average price of fifty dollars after 1992, he was entitled to another $6.6 million for every additional quarter he stayed on the job (in addition to another stock award of $8.8 million).

Almost as an afterthought, Iacocca held out for one more benefit. Since Greenwald's departure upset his somewhat hazy retirement plans, Chrysler had to agree to buy two of his homes, valued at a total of nearly $1.7 million, in Palm Beach and in Bloomfield Hills, north of Detroit.

Jerry Greenwald's fears about the viability of his new job proved wise. By autumn of 1990, the United Air Lines deal had failed. Greenwald had labored long and hard during the summer. No one had appreciated how terrible the balance sheets at many of the world's major banks looked just then. The Japanese banks were overextended too and had little appetite for lending $4.5 billion to United's unions. He tried his best but wasn't able to secure financing. Greenwald was paid his $9 million, as agreed. Soon after, he joined Dillon Read as a managing director to work on investment banking deals. Nine million bucks for ninety day's work! From Detroit to Wall Street, heads shook in disapproval and envy.

A month or so later, *Playboy* published an interview in which Iacocca savaged Greenwald for leaving Chrysler. According to Iacocca's interpretation of recent history, he had been planning to announce his retirement and appoint a new management on November 1. Greenwald crossed him up by quitting. Strangely, he had never mentioned that scenario to Greenwald when the two of them talked in Italy.

> **Playboy:** Do you fault him for leaving?
> **Iacocca:** Sure I fault him. Why not? He grew up with me. He'd been in the car business for thirty-two years, same two companies as me—Ford and Chrysler. Isn't there anything sacred anymore? Isn't there any loyalty to anything? I told him, "Jerry, it's the Nineties. The Eighties were this kind of thing; you should have done it then, and I would have written you off as caught up in the Yuppie movement. But that's over. The mere fact that they can pay nine million dollars for ninety days shows that it's go-go time again. After you've drawn your nine million dollars—and even if you become C.E.O.—you'll still look back on all the friends you talked into coming with you to Chrysler, and it's *still* an act of walking out on the gang. Easy come, easy go." I even told him, "If you want to climb a mountain twice, do it with Chrysler."

To read in a national magazine that Iacocca villified him for leaving Chrysler stung Jerry Greenwald—particularly since he had come to believe that Iacocca never really intended to leave and had manipulated Greenwald's exit to his own advantage with the board of directors. And calling Greenwald a Yuppie and a holdover from the 1980s was particularly offensive, especially since Iacocca had paid himself more than any other auto executive in history ever got. But that was Lee: self-centered, greedy, quick to blame others, and blind to his own faults. And when you crossed him, look out.

Throughout 1991, Greenwald worked at Dillon Read, never quite abandoning hope that he would someday regain the limelight, perhaps with a company that needed a chief executive officer. He raised his hand to volunteer to clean up the savings-and-loan mess as head of the Resolution Trust Corporation but was passed over in favor of Albert Casey.

For about six months following the publication of the *Playboy* interview, Greenwald and Iacocca didn't speak. Glenda Greenwald and Darrien Earle—soon to be Iacocca's third wife—stayed in touch, however. Each suggested to her mate that, despite the harsh words, the two executives had been close allies for a long time and ought to bury the hatchet. There ensued a somewhat cool rapprochement between Iacocca and Greenwald. Iacocca even consulted with Greenwald, in his capacity as an investment banker for Dillon Read, about Chrysler's search for a successor, a search that Jerry Greenwald privately believed would never bear any fruit.

Meanwhile, an interview appeared in *USA Today* in which Iacocca was asked about Greenwald's departure. He did what he had to do, Iacocca said, and we're friends. The statement was his public declaration that the hostility between the two men was forgotten. Iacocca still needed Jerry Greenwald for reasons that would soon be apparent.

GODZILLA
ECONOMICS

THE FIRST RAYS of sun peaked over the San Gabriel Mountains, spreading light upon the awakening city of Los Angeles. Mist hovered above the surf rolling gently ashore in Palos Verdes, Malibu, and the communities arrayed along the nation's West Coast.

The morning still was broken in a quintessentially American way, by the ignition of hundreds of thousands of automobile engines. At 7:00 A.M. the freeways were humming, and within an hour traffic filled every thoroughfare. From a bridge crossing the San Diego Freeway, the automobiles described an unbroken stream of metal and glass, gleaming arteries surging in every direction.

The grocers, the lawyers, the manicurists—they drove two hours and more from stucco homes in the San Fernando Valley and Orange County to offices and parlors sprinkled along the banks of this great automotive river. Seated behind their steering wheels as traffic inched forward, they chatted on the phone, hummed along with the radio, glanced at their newspapers; they often felt as though their vehicles were extensions of themselves—which is not surprising since they spent more waking time in their cars than at home.

Observing each other from behind the wheel, Californians regarded their cars as essential items of clothing, statements about themselves, defining marks of status. Here you wore—not just drove—your car. Strangers at cocktail parties broke the ice by asking other strangers what kind of car they drove. Choosing a car was like picking a spouse or, at least, a date. Falling in love with a car meant defending it, babying it, keeping it shiny. For many, no mere car wash would do. For a few hundred dollars a "detailer" would come to your garage and swab your car lovingly with nonscratching cloth, removing dust from every last crevice with Q-tips.

In the century since cars began appearing in America, California has evolved into the car-craziest spot in a car-crazy country, a place where personal transportation is regarded as elemental as food, water, or shelter. California is a coastal culture, where strange new ideas wash up and are tested before advancing to the hinterland. The culture that produced Disneyland, Ronald Reagan, hot tubs, and the Apple computer is the one where cars from Japan first began to challenge the Cadillacs, Chevys, Fords, and Dodges that once reigned supreme. Beyond California's convenient geographic location, Japanese carmakers knew that Californians are more willing than most other Americans to try something a little different. In the late 1950s and early 1960s only a very few car buyers were willing to entertain the notion of compact, lighter automobiles exotically colored in turquoise and mango.

The cars that Toyota and Nissan introduced in California thirty years ago suggested very little of their ultimate impact on economic events; they were small and flimsy, a species inferior in many ways to what most Americans were driving. In the early years executives of the U.S. automobile industry treated Japanese cars with curiosity and contempt. Little shitboxes or beer cans, Detroit called them. Their spare frames and tiny engines, humming like sewing machines, more properly belonged in the narrow streets and parking spaces of Japan. Yet year by year, as they grew in size and power, they became harder to dismiss. The engines and transmissions became more sophisticated. Low price—made possible in part by the cheap yen—and superior fuel efficiency appealed to conservation-minded buyers, particularly after the oil shocks of the 1970s.

Japanese automotive design embodied an undeniable cleverness and originality. Most American automakers, however, couldn't see it. They could not fathom the logic of people who bought the Japanese cars. Just as silly to them were the drivers who had bought the homely, noisy little Beetle from West Germany. But by the late 1970s the proliferation of Hondas and Toyotas had begun to unsettle Detroit, planting in the industry's consciousness the seeds of apprehension. The more perceptive minds of the U.S. auto industry sensed that the threat from the Far East had turned serious. Even so, Detroit's collective wisdom refused to acknowledge the appeal of Japanese cars, or to try to learn how Japanese automakers had accomplished their considerable feats and match them with vehicles that attracted customers in this new way.

Instead, the leaders of the U.S. auto industry blamed their troubles on others: U.S. government policy, the United Auto Workers Union, and Japan. The industry blamed the U.S. government because, in Detroit's view, Washington heaped too many regulations on automakers and failed to pursue economic policies (trade barriers) that would restrain foreign automakers in the United States. Auto executives blamed the UAW for failing to cooperate adequately with initiatives to cut costs and raise productivity to the level of Japanese factories. Finally, Japan itself was targeted directly by automakers because that government and its automakers supposedly conspired to pursue predatory trade and economic policies designed to destroy the U.S. industry.

Detroit—the name that represented a way of life for hundreds of thousands of assembly-line workers, executives, designers, suppliers, and shippers across the U.S. auto industry—seemed helpless to comprehend clearly what threatened it and, therefore, to devise a defense and counterattack. The industry's best brains were paralyzed by the welter of legends and half-truths they had created to explain their humiliating loss of power and preeminence. The common element in these scare stories was: Japan. Essentially America was being told that Detroit—and, by extension, the nation's economy—was under attack by an all-powerful, supernatural enemy. Japan, in the form of a Godzilla-like monster, had suddenly emerged from the Pacific, reared on its mighty haunches, and, without warning, set about trashing our nation's auto industry.

Detroit appeared powerless to do anything but hope for a savior like Dr. Serizawa, the hero of the 1954 Japanese monster movie. Since the Godzilla that threatened Japan had been impervious to conventional weaponry, Serizawa invented an "oxygen destroyer" to drive the beast back to its suboceanic lair. Detroit wanted Washington to invent an "oxygen destroyer" for it, some defensive weapon in the form of a tariff or other trade barrier. American administrations have long been ambivalent about protecting the U.S. auto industry from foreign competition, partly because of ideological commitments to free trade and partly because Detroit never mounted a very strong case for defense by the government. Indeed, until recent years Detroit remained ignorant of the forces it faced. Though Americans had occupied Japan after World War II and have since visited the country in great numbers, American auto executives, with few exceptions, knew little about the country's history, customs, or the nature of its technological advantages.

To argue that Detroit was diabolically lulled by Japan into complacency would be self-serving. A key function of well-paid corporate executives is to assess competitive threats, particularly when they are in an embryonic stage and easier to confront. And this is precisely what Detroit failed to do. When the 1958 Toyota Crown, the first Japanese model to be imported into this country, arrived here, it was already popular in Japan. Here it was seen as a car that vibrated badly at sixty miles per hour and overheated when driven over hilly terrain. Its low price wasn't too startling in light of the weakness of the Japanese currency, Detroit thought, and, in any case, the Crown wasn't expected to inspire many drivers to trade in their Ford Fairlanes or Chevy Bel Airs. Sales were in the hundreds, not even a blip on Detroit's radar screen, and certainly not worth worry for a country that bought ten million new cars and trucks annually.

Because Detroit paid little attention to Japan, it was completely unaware that the Japanese automobile industry of the early 1960s stood on the threshold of breathtaking engineering, manufacturing, and organizational innovations. With breakthroughs in design and quality came an appreciation by American customers for whom the legend "Made in Japan" meant tinny prizes found in Crackerjack boxes. Detroit only had to take one hard look at the rising line on Japan's U.S. sales charts to realize that that Asian nation's achievements deserved deep study. Japan's gross exports rose from $13 billion in 1968 to $37 billion in 1973 and $147 billion in 1983. Consumers, especially in the United States, increasingly welcomed its merchandise and were eager to buy its $10,000 cars, not just the $100 Sony Walkmen.

As Japan's exports to the United States grew, their collateral impact on U.S. economic and community life became more and more profound. The sales volume of American companies that competed with Japan dropped, as did the price that some American goods commanded. Pressure to reduce costs often took its toll in quality and innovation. Increasingly, automotive factories were closing as drivers preferred what Japan was selling to what Detroit had to offer. The business pages of America's newspapers read more like obituary pages as they reported the closings of factory after doomed factory: Mahwah, New Jersey; Norwood, Ohio; Flint, Michigan; Kenosha, Wisconsin. Each shut-down factory pushed thousands onto unemployment rolls, expunged taxable property from the tax rolls, reduced the revenue of cities and school boards. Not surprisingly, the clamor for remedies grew from unions, local politicians, community organizers, and industry itself.

Since the 1920s a vibrant automotive culture has been the sine qua non of strong national economies worldwide, and that is the reason why aspiring, ambitious nations, from Mexico to Thailand, want to build cars. Automotive manufacturing requires steel, glass, and rubber manufacturing, as well as related high-skill industries like tool-and-die-making and capital equipment. But above all, automaking requires highly educated, competent engineers and managers, who, if they succeed, become the exemplars and progenitors of a nation's executive talent. In the United States, for example, former trainees from Ford, General Motors, and Chrysler have migrated to the leadership ranks of many nonautomotive companies.

So it's hardly surprising when a government swings into action to save a troubled automaker. As the Chrysler Corporation teetered on the brink of failure in 1979, the United States was faced with a decision whether to let a bankruptcy judge supervise the reorganization of Chrysler or do the job itself. Thousands of troubled companies had floundered in the recession and energy crisis precipitated by the fall of the shah of Iran. But no casualty was as large or as important economically to so many as Chrysler. The efforts of the Treasury and the Federal Reserve to guarantee loans and coax concessions from labor, bankers, and suppliers testified to the importance the nation attached to its auto industry and to its major employers.

Chrysler's need for a federal bailout can't be explained adequately by a weak economy alone. Nor was Chrysler in trouble because of the cost of regulation. Chrysler landed in the soup mostly due to the spectacularly awful cars it built during the 1970s, which drove customers to Japanese and other brands. The energy crises of 1973 and 1979, by making fuel-efficient cars more desirable than gas-guzzlers, only aggravated consumer disenchantment with poor quality. Chrysler was heavily committed to bigger cars with inefficient rear-wheel-drive. But Chrysler's cars weren't merely out of phase with trends in fashion or fuel economy; they were shoddily built, as disgruntled owners of the defect-plagued Dodge Aspen and Plymouth Volare have verified. (Manufacturing quality wasn't much better at other U.S. assembly plants; but as the smallest of the Big Three, Chrysler's had a smaller financial cushion and a weaker network of retail dealers to contend with defects and poor workmanship.)

The government agonized over whether to rescue Chrysler without understanding that the rest of the industry was not in much better shape. Unfortunately, Chrysler's woes did not draw enough attention to the entire industry's worsening relations with its customers. Americans loved cars. Every autumn they waited for the sweet, corny ritual of new-model introduction. Car owners were excited by new body styles, features like stereo and air conditioning, powerful engines. They hummed Beach Boys tunes about racing hot rods to the beach. They invented reasons to sign up for forty-eight months of new-car payments, however unaffordable.

As car prices rose over the years and dealers took the steady stream of customers for granted, the once exciting experience of purchasing a new car degenerated into a series of hassles and disappointments. Customers regarded negotiating for a new car with only slightly more enthusiasm than submitting to root-canal surgery. Cars didn't arrive when promised, or they weren't equipped according to the agreement. Quality was often poor. Why in the world, consumers wondered, should a new car shed parts or begin to rattle on the way home from the dealer? Why did paint peel or metal rust after only a few months of service? Since automakers honored their warranty claims only grudgingly, dealers had little incentive to take care of customers. Customers, in turn, were forced to grovel, scream, or threaten legal action to get dealers to repair defects. Worst of all, new-car buyers lived permanently in the shadow of humiliation and dread, worrying when the next breakdown would occur, wondering how Detroit had duped him or her into buying another domestic lemon while friends were raving about their new Hondas.

Contemplating a switch to a Japanese car provoked a painful ambivalence for many Americans. Buying a Japanese car meant, in a sense, breaking the faith. Buying a Japanese car was casting a ballot against America, not just against Detroit. The tide of Japanese cars forced factories to close, workers to lose jobs, dealerships to fail. Workers, families, the nation had stood tall against Japanese aggression in the Pacific during World War II, never dreaming that one day they might be forced into economic combat with Japan at home.

The ripples of economic unrest caused by the tide of Japanese cars extended outward — to the supermarket, small shops and professional practices, state and local government. Laid-off workers needed unemployment compensation to survive, which necessitated higher taxes. The rising trade deficit inexorably eroded the value of the American dollar, a basic measure of national prosperity. Though California was the spiritual nexus of the automotive culture, the economic impact of Japanese cars hit the Midwest most severely, especially southeastern Michigan. A down year for the automakers instantly translated into hard times for the clothing stores on Woodward Avenue in Detroit, the dentists in Bay City, and the luncheonettes around the factories in Sterling Heights.

In Detroit and other factory towns, the sagging fortunes of the American auto industry dealt an especially tough blow to those who migrated from the South in search of better-paying jobs in the 1930s through the 1950s. These undereducated sons and daughters of sharecroppers had been lifted from near-poverty to healthy middle-class status by the auto industry. For many autoworkers, retraining was unrealistic without remedial basic education; and so, for many the shutdown of an auto plant spelled an instantaneous return to poverty.

Worried about economic consequences, some Americans refused to consider Japanese cars. Others decided to select the best vehicle for the best price, keeping their own economic well-being foremost in mind. In the spirit of free-market capitalism, some reasoned that second-rate automakers, the builders of lemons, needed an incentive to build better cars; buying Japanese provided the incentive. Besides, the automakers themselves had become aggressive buyers of Japanese parts and engines. If automakers were willing to threaten the health of U.S. suppliers for the sake of better quality and value, why should car owners behave any differently?

For the consumer, economic trade-offs were never clear-cut when patriotism became a factor. A consumer who says, "All things being equal, I want to buy an American car," expresses a sentiment that is safe and easy. Consumers know, of course, that all things seldom are equal. No one wanted to put fellow Americans out of work, nor did they wish to waste thousands of dollars on a substandard product. But years of buying cars that were "good enough" rather than the very best meant consumers unwittingly had encouraged practices that dragged the U.S. auto industry to its parlous state.

Chrysler's imminent bankruptcy brought these economic trade-offs into bold relief. Many in government and academia favored letting Chrysler fail, arguing that a free-market economy should not perpetuate poorly managed enterprise. Propping up failing companies only forestalled the inevitable. If government decided which companies should live and which should die, the day was fast approaching when companies would need a government license and bureaucrats would decide what was to be produced. A brief visit to

Havana or Minsk showed how well the philosophy of central economic planning was working.

Perhaps, on the other hand, Chrysler was a harbinger of worse economic troubles on the horizon. Perhaps the entire U.S. auto industry was headed for the same fate as industries like consumer electronics and steel unless the government saved Chrysler. The bailout was an opportunity to strengthen the quality of business leadership with the annealing fires of competition.

Strangely, the men of Detroit (and they *were* virtually all men) weren't apologetic about the decline of their industry. Give us the same conditions as the Japanese, they demanded, and we will quickly shove them back where they came from. Japan has cheap labor, a low cost for capital, and a government that helps its companies. Give us those advantages too. Give us "a level playing field"—a cliché mercilessly overused by corporate speech writers—and we can compete with anyone. The battle cry was directed at government. Detroit, in short, wanted protection cloaked in the guise of fair play.

Publicly and privately, auto executives flayed Japan for flooding the U.S. market with imports. If Japan wanted to sell cars here, they argued, their automakers should build them here. Japanese automakers, they believed, would be unable to outperform the Big Three in America's backyard. With the unions and government regulators hanging on their backs, the Japanese couldn't do any better than the Big Three, or so the argument went.

The Reagan administration, committed to free trade but politically sensitive to auto industry unrest, sponsored a "voluntary" restraint agreement in 1981 under which Japan agreed to limit car exports to the United States, a move that would buy time for Detroit. With competition restrained, the U.S. industry could spend more on modernizing plants, creating new models, and improving technology. American car buyers, meanwhile, were paying for the artificial scarcity of Japanese cars in the form of higher prices. And Japanese automakers and their U.S. dealers were reaping hefty profits since demand for their cars far outstripped supply.

One important outcome of voluntary restraints was unexpected by Detroit and quite unpleasant. Recognizing that exports were an irritant to relations with the United States, Japanese automakers and suppliers began a long-term strategy of building plants in the United States. Their motivation was not fair play, of course, but an understanding that political tension had to be defused. A second reason for building plants in the United States was to shorten the lines of supply and communication between factory and customer, an important tenet of Japanese manufacturing doctrine.

By the late 1980s, Detroit realized that it needed *more* protection. As chairman of the automaker with the most to lose from Japanese imports, Lee Iacocca had the greatest incentive among auto executives to push for government intervention. Chrysler was the weakest financially of the Big Three. Following the government bailout, it built mostly fuel-efficient, basic models that competed directly against Japanese models. General Motors, as a matter of corporate policy, favored free trade. Ford executives, though they grumbled, didn't often speak out against Japan. They thought complaints made Ford sound whiny.

The role of field general in the ideological battle against Japan suited Iacocca's reputation as a master communicator who understood public tastes and sentiments. From his first day at Chrysler in 1979, Iacocca almost immediately found himself in Washington pleading for federal aid. From that time through 1983, Chrysler was virtually a ward of the state, operating under loose government supervision with little money to invest in new models or the expensive new technology required by federal regulations.

Iacocca understood how poorly equipped Chrysler was to battle competitors in the marketplace. He had to hold off Japan until he could improve Chrysler's arsenal. Initially, Japanese competition wasn't Iacocca's biggest headache. His first task was convincing the government to provide fresh financing, something banks were reluctant to do. A conventional bankruptcy, he argued, would destroy confidence in the automaker and force dealers into bankruptcy, a catastrophe from which he believed there could be no recovery.

The government declined to guarantee new Chrysler loans unless the automaker first received concessions from suppliers, creditors, and the United Auto Workers Union. In this way the government was forcing systemic change on the automaker's operations. Chrysler wasn't allowed to buy parts, borrow money, or negotiate labor contracts in the traditional, outmoded manner. The government also thought the style of Chrysler's executives needed reform. As president of Ford, Iacocca had grown accustomed to imperial treatment—lavish parties, corporate jets, suites at the Waldorf, weekends at resorts, all on the company tab. Auto executives may have treated themselves royally when they were the masters of their own realm, but the taxpayers weren't about to subsidize a royal budget.

"Quite frankly, I would rather not be here at all," Iacocca told the House Banking Committee in 1979. "I am a strong advocate of the free-enterprise system. I grew up in it and slugged my way through it for thirty-three years. I am sure you share my conviction that in the long run the answers to our problems are going to be found not in the halls of Congress, but in the marketplace."

Iacocca had little reason to suspect how very much more difficult the marketplace was about to become for Chrysler and other American automakers. In 1982, only two years after the bailout, Chrysler's finances were stabilizing. By 1983 the company was able to repay the guaranteed loans early, thus removing the yoke of government supervision. Chrysler's car models, smaller and utilitarian, were well suited to consumers who remembered gas lines. The national economy was recovering, which drew car buyers back into showrooms and helped replenish Chrysler's coffers. Japanese automakers, though restrained by the voluntary agreement, nevertheless were winning over customers with small, fuel-efficient models like the Toyota Corolla and Honda Civic. When the Japanese automakers' U.S. plants opened, their output was aimed like a dagger at Chrysler's heart. The Japanese, moreover, were also preparing to introduce high-profit luxury models to compete with Cadillac, Lincoln, and other U.S. luxury brands.

Iacocca, with an eye on the growing Japanese presence, stepped up his call for stronger measures. The name he gave to what he was promoting was neither "protectionism" nor "free trade" but "fair trade." By Iacocca's logic, if Japan posed barriers to U.S. goods, the United States should retaliate with barriers. He also accused Japan of manipulating its currency to keep the yen artificially weak, thereby making Japanese exports cheaper and more attractive to foreign consumers.

Iacocca's position had some popular appeal, particularly among those who disliked or mistrusted Japan, remembering it as an enemy in World War II. Many Americans felt visceral stress when they saw Japanese cars. But when the U.S. Treasury asked Iacocca to substantiate accusations of unfair trade practices in his industry, he couldn't. He cited the well known examples of how the Japanese had closed their markets to U.S.-made pharmaceuticals and optics. Initially, he was careful not to mention automobiles, because he knew that no U.S. automaker to that point, including Chrysler, had ever mounted a serious effort to sell cars in Japan.

Selling the idea of trade retaliation to the entire American public was a dicey proposition. Since World War II every administration has preached liberal trading practices to the underdeveloped world as the key to freedom and prosperity. With the Smoot-Hawley tariff and the economic collapse of the 1930s fresh in mind, the General Agreement on Tariffs and Trade (GATT), which was aimed at gradually reducing tariffs and other barriers worldwide, was negotiated and signed by the Allied nations in 1947. As tariffs and barriers fell, trade increased, adding to prosperity among trading partners, all in accordance with free-market economic theory.

Critics have argued, with some justification, that nations such as the United States and Great Britain tend to promote free trade most stridently when they're strong and certain that their own industries will be dominant. Thus, the United States worried little about its auto industry before Chrysler's brush with bankruptcy. In promoting free trade after World War II, the United States had an additional goal: forging economic and political alliances between industrialized nations—and against the Soviet Union. The U.S. policy of containing Communist influence undoubtedly was served by strengthening trade among non-Communist nations.

As the economies of emerging nations began to catch up to the United States in the 1970s, free trade ideology collided with the narrow interests of politicians whose local districts were feeling the heat of competition. Overnight, American industries were besieged by low-price imports. The importation of steel into the United States was a particular sore point. America once had a lock on steelmaking, automaking, and a host of other industries; now others had the key.

American companies put pressure on local politicians, who filed complaints under U.S. antidumping laws. Dumping is typically defined as selling below the cost of production or below what a product sells for in its home market. But violations are hard to prove, since definitions of what constitutes cost and government subsidy differ from country to country. Most free-trade theorists scorn antidumping laws as nonsensical, since, they argue, no one can stay in business long selling at a loss or as a ward of the state. In many instances, laws against dumping are used as quasi-legal method of holding down competition from abroad. Filing a dumping complaint with little expense to the complainers is simple; proving the charge is quite another matter. The litigation, however, can tie up a foreign competitor for years. Comparing production costs in two countries is an arcane exercise. Taxes, for example, may be computed on a wholly different basis. Compensation packages for workers don't necessarily include the same benefits. Currencies fluctuate. The comparison of selling prices in two different countries isn't straightforward either. Dumping complaints are, however, quite a useful means of intimidating foreign competitors.

The rhetoric about "unfair" trade practices often missed the essential point of why consumers favored Japanese brands. In car buying, the selling price often isn't the deciding factor for the buyer. The Big Three automakers knew this fact from their research but brushed it aside when arguing for trade sanctions against Japan. For commodities like wheat, steel, or semiconductor chips, price is important. Wheat, more or less, is wheat. Iacocca knew better than anyone that buying a car is a complex emotional decision that encompasses many factors besides price. A customer gravitates toward a minivan, say, because he likes the way it looks, because it fits his need for more room, because it is reliable, or because the dealer has a reputation for trustworthiness. Likewise, during the energy panic of 1979 and

1980, buyers were suddenly willing to pay inordinate sums for sparsely equipped, underpowered cars that were fuel-efficient.

The agreement to limit exports from Japan distorted the economics of car buying in several ways. Demand for Japanese cars was already rising, and since their allotment was limited, Japanese automakers wisely decided to export larger, better-equipped, more expensive models that would yield them more profit. With demand exceeding supply, the Japanese also were able to raise prices, further maximizing profit per unit. This, in turn, afforded the U.S. industry a pricing "umbrella" under which it could raise its own prices. Higher prices on Japanese and American cars constituted a "surcharge" on consumers estimated at $5 billion annually—paid directly to automakers and dealers.

Limiting Japanese exports was a bonanza for some U.S. dealers of Japanese brands, and a disaster for others. Automobile dealerships, with few exceptions, are independently owned enterprises that operate under franchise agreements with auto manufacturers. When a car is popular, the dealer normally prospers, and vice versa. Dealers that secured healthy allocations of Hondas and Toyotas in the early 1980s enjoyed the patronage of buyers who were waiting in line to buy Japanese cars. Dealers who couldn't get enough cars had trouble staying in business. (Unscrupulous dealers sometimes bribed executives in charge of distribution with gifts and cash in order to receive more cars than they were entitled to. In one spectacular case, several Honda sales executives were indicted in 1994 and charged with racketeering and mail fraud in connection with bribes from dealers.)

For dealers shrewd or lucky enough to hold the best franchises there was little trick to selling Japanese cars; they sold themselves. Voluntary restraints made an already lucrative business all the more juicy. Though the dealers were legally entitled to charge whatever they wished, Japanese automakers discouraged them from jacking up their prices. But even without raising prices, dealers knew how to "pack" cars with pinstriping, rustproofing, and other high-profit, unnecessary options.

Consumer ambivalence over American cars was difficult for

Detroit to swallow. Iacocca and his contemporaries were men who had led the American auto industry to its glory days. College graduates of the 1950s and 1960s, these auto executives regarded themselves as the elite of American business, the cream of the crop. They had inherited an affluent society from the generation that won World War II, and they distinguished themselves by building the biggest corporations on earth. Long before *teamwork* and *cooperation* became the buzzwords of industry, the individualists who occupied choice corner offices on high floors were winning their positions in ruthless corporate combat with one another. Being a team player was a euphemism for going along with the boss. Careers were made by attaching oneself to the right boss and making him look good. Self-doubt was rare for this breed, humility rarer still. Indeed, the first rules of corporate infighting in Detroit were never admit failure, never back down, and never concede weakness. This mindset, so effective in forging the culture that brought automobility to a mass culture, was not well suited for responding to strange and formidable adversaries.

The executive culture exemplified by Iacocca and his ilk shaped the automotive industry, brooking little criticism from anyone, until the 1970s. The bosses of the auto industry decided everything—how big, how powerful, how safe, what color, which amenities—and they entertained very few ideas from consumers or regulators, much less from upstarts in places like Japan and Germany.

The advent of so simple and obvious a safety feature as the seatbelt was postponed for years because Detroit executives stonewalled it. General Motors engineers testified in Washington to the seatbelt's uselessness. Automakers knew plenty about safety, far more than the government or safety advocates. But they used their knowledge, and the self-assurance that they knew what was best for customers, to fight safety legislation until they believed the financial payoff outweighed the cost. Whenever consumer advocates lobbied for safer, more efficient, environmentally cleaner, and less expensive cars, the industry went straight to the barricades. Who are these reformers, automakers wanted to know. Who represents them and on what authority do they speak on behalf of "our customers"?

When finally forced to respond to the emerging clout of consumers, automakers produced an impressive array of safety and pollution control devices. In recent years the automakers have introduced catalytic converters, air bags, antilock brakes, and clever safety features like integrated child-safety seats—something Chrysler introduced before consumers even imagined it possible.

Likewise in the competitive arena, Detroit delayed confronting Japanese cars until it could delay no longer. By the time California's enthusiasm had spread to the East Coast, across the Sunbelt, and into the Midwest, Lee Iacocca and like-minded executives were caught flat-footed. All that was left to do was demonize Japan as a modern-day Godzilla and cry for government protection.

Iacocca's response to Japan characterized the air of defeatism that ran through Chrysler in the 1980s. The strategy of scapegoating Japan seemed so desperate at times; Iacocca's statements made one wonder whether he was spurred by fear of Chrysler's demise.

NAGOYA RULES

OWNERS OF FORDS or Chevys or Dodges who decided to test-drive a Honda Accord or a Toyota Corolla in the early 1980s were often surprised by the immediate sensation that this new machine represented an unanticipated and pleasing leap in automotive technology. The exact feeling was not easy to articulate. The Japanese models, though a bit smaller, seemed to be constructed of more or less the same materials as domestic models. The controls and gauges were located in roughly the same place. Perhaps the craftsmanship was a bit better. Perhaps the Japanese models sounded a bit quieter, had a distinctive shape, felt a little tighter, rode a little smoother.

If the differences between Japanese and American cars were difficult to quantify or explain, they were no less real. The impressions, though slightly intangible, were the result of dramatic improvements in design and manufacturing technology. Until Jim Womack and the MIT researchers analyzed and explained the advances that vaulted Japan ahead in their book *The Machine That Changed the World,* what precisely was happening inside Japanese car factories was poorly understood in the West except by a handful of experts.

According to the arguments of Lee Iacocca and many leaders of the U.S. industry, Japan's automaking prowess rested on the efforts of workers fanatically dedicated to quality, as well as on the support of a government that kept foreign products out of its own country while helping Japanese automakers conquer markets overseas. Although

Iacocca professed to admire what he viewed as a coordinated strategy by Japanese business and government, he attacked it as wrong and unfair because, by his reckoning, Japan's automakers were spared having to build cars in the United States, where high health-care costs, a tough union, less diligent workers, and unsympathetic government regulators made life difficult for automakers.

To some extent Iacocca was correct. Japanese factory workers dedicated a great deal of time and diligence to their tasks, as measured against American standards, though this dedication was not the only or even the primary reason for the success of Toyota, Honda, and others. True, Japanese workers often worked overtime without complaint and threw themselves into their jobs with enthusiasm; but the significance of the myth of the overachieving Japanese (and, by implication, the underachieving American) autoworkers was not put into its true perspective until American workers employed by Honda, Toyota, and Nissan began building cars in the United States that matched in quality those built in Japan.

By the mid-1980s, Honda was assembling cars in Marysville and Liberty, Ohio; Nissan was building cars and trucks in Smyrna, Tennessee; and Toyota was building cars in Fremont, California, in a joint venture with General Motors, and on its own in Georgetown, Kentucky—with American workers using the same methods as those in the Japanese factories. None of the workforces except that at the Fremont plant was represented by the United Auto Workers union.

The ability of "transplant" factories to produce high-quality vehicles debunked the myth that the manufacturing prowess of Japanese companies somehow stemmed from ethnic or cultural roots. If Japanese dedication to work was the deciding issue, the quality among Japanese automakers wouldn't differ so sharply from company to company. Nor would American workers be able to function so well in factories like Honda's Marysville plant, which produced the Honda Accord.

Further undercutting ethnic stereotyping, not every Japanese automaker found itself able to transfer production smoothly to the United States. For example, Mazda's workforce at its Flat Rock, Michigan, factory adapted slowly and with difficulty to the automaker's quicker, more demanding pace of production. Tension arose between Mazda's management and the United Auto Workers union, which represented Mazda workers. Workers complained that the pace was too fast and that supervisors ignored medical and safety problems. Japanese executives were disappointed that workers didn't embrace their jobs with the same spirit as their Japanese counterparts.

Joseph J. Fucini and Suzy Fucini examined relations among Mazda managers, the UAW, and the rank and file in their book *Working for the Japanese* and concluded that the Flat Rock plant fell far short of what Mazda had hoped to accomplish in the field of labor relations. The title of the Fucinis' book was unfortunate because it suggested that the problems Mazda encountered in its Flat Rock plant were typical of any meeting of Japanese management and American labor. In actuality, the events at Flat Rock represented the particular experiences of a single automaker and contrasted sharply with the American manufacturing experiences of Honda, Toyota, and Nissan.

Manufacturing practices developed in Japanese factories seemed, in concept, simple enough to Western minds. Workers could understand them easily. Taken individually, no single practice seemed terribly profound. Together they appeared to be more or less a collection of commonsense notions, such as always conserving space by moving machines closer to one another, or color-coding tools to avoid misplacing them. The philosophy of the Japanese factory relied heavily on the eradication of *muda* (waste) in all its forms: material, effort, and time. Manufacturers also stressed the concept of *kaizen* (continuous improvement). Japanese workers were exhorted to find every glitch and defect in the manufacturing process and fix them, never to accept the status quo. Fewer and fewer defects and less waste were benefits that flowed quite naturally from a dedication to correcting and improving everything possible.

Stamping out *muda* and pursuing quality through *kaizen* were concepts that applied beyond the factory. They belonged to a dynamic framework that connected assembly to distribution, product design, marketing, parts supply, and every other aspect of ultimately satisfying customers. Thousands of little efficiencies, practiced across the corporation, resulted in enormous increases in productivity and decreases in the number of customer complaints. American factory managers, upon hearing about the productivity of Japanese factories, often sneered that the Japanese weren't doing anything that hadn't been thought of previously by W. Edwards Deming and other Americans. Deming, for example, implored American manufacturers to learn statistical process control (SPC), a way of measuring and monitoring the tiny deviations from manufacturing tolerances that caused the poor fit of parts.

Those who sneered were right. The ideas embraced by Japanese manufacturers *had* been thought of long ago. That was the point. The Japanese had learned a great deal of what American manufacturers had forgotten or ignored, and they turned that knowledge to their advantage. The millions of little improvements and methods for reducing waste applied to far more than automaking; together they constituted a way of making any number of things for the mass market. The MIT researchers labeled the new system, for lack of a better name, "lean manufacturing." As far as Jim Womack and his colleagues were concerned, the advent of lean manufacturing spelled no less than the obsolescence of classic mass manufacturing. Toyota's cars were not the triumph of Japan, or of Japanese culture, or of one auto industry over another, but the replacement of one era of industrial history by another.

The two men credited with developing this new way to make cars — to make and sell anything, really — were Eiji Toyoda and Taiichi Ohno of the Toyota Motor Corporation. To study the best of Japanese automotive manufacturing is to study the Toyota Production System, which evolved from the 1950s through the 1970s, under the guidance of these two Toyota executives, to become the standard for the Japanese industry. Ohno, Toyota's production chief, introduced practical methods for saving time and material and later explained his ideas in a book, *Toyota Production System*, which was translated into English. Toyoda, as a managing director and member of the automaker's founding family, had visited Ford's Rouge manufacturing complex near Detroit in 1950. The Rouge facility integrated steelmaking, glassmaking, parts manufacture, and final assembly. Iron ore and other raw material were delivered to the Rouge's docks.

At the other end of the complex, cars rolled off the assembly line. Toyoda was impressed with the massive scale of Rouge and its integration. At a glance he could see the endless opportunities for improvements and efficiency. His visit to the plant filled Toyoda with hope that the Toyota Motor Corporation one day might match Ford's technical ability, if not its scale of production.

Of course, Toyota never could have developed the Toyota Production System, the first great demonstration of lean manufacturing, without first understanding mass manufacturing. Though Henry Ford is regarded by some as the inventor of the car, the first models actually were invented in Europe twenty years before he built his first car. Ford's actual contribution to industrial history was more important. Ford, in 1908, adapted the moving assembly line for auto production, which arrived hand-in-hand with another key Ford innovation: interchangeable parts. Previously, cars had been built by hand, one by one, from parts constructed individually by craftsmen. Two cars may have looked similar but many parts from one would not have fit another without cutting or bending. The introduction of the assembly line and interchangeable parts permitted relatively low-skilled workers to stand all day in one spot with a stack of identical parts and easily attach them, one by one, as the cars rolled by.

In factories modeled after Ford's first automated plant in Highland Park, Michigan, all manual labor was divided into the most simple, repetitive tasks. This method of organizing the assembly line borrowed from the thinking of Frederick W. Taylor, an efficiency expert whose ideas were then gaining popularity. Workers needed few if any skills because their responsibilities were limited. Many Ford workers didn't speak English and didn't need to. The work of skilled laborers and engineers was more complex, but their jobs too were carved into small portions: one employee ordered parts, another fixed machines, someone else inspected for quality. Usually the gruffest, burliest man on the payroll was appointed foreman, to enforce discipline and make sure the assembly line kept rolling. *Taylorism* became the intellect's shorthand to describe the chopping of work into smaller and smaller tasks.

To make cars more popular and available to the common man, Henry Ford constantly searched for ways to increase the speed of production. He designed parts for ease of assembly; he dedicated machines to specific parts so that they could be operated almost continuously by workers with minimal skills. When a new part was needed, the old machine was often junked; Henry Ford deemed this practice preferable to buying machines of more general design that could produce several different parts but had to be tended by a skilled tradesman. Just as most Ford workers were assigned a narrowly defined task, there was just one principal decision-maker, Henry Ford himself.

As the variety of cars evolved to suit many tastes and purposes, tension arose between the need to standardize parts and tools as much as possible to keep costs down and the proliferation of different models. Suddenly, the financial stakes of decisions were higher and the circumstances affecting them far more complex. Professional managers with knowledge of finance, marketing, and engineering were needed who could analyze the issues more rigorously.

Alfred P. Sloan, Jr., who rescued and reorganized the General Motors Corporation, conceived the model of a corporation that left decision-making to autonomous divisions while leaving overall strategy to corporate executives. Under Sloan, there arose a competent managerial class with specialized skills. Sloan's organizational brilliance was validated by GM's rapid growth; in the late 1920s it outpaced Ford, which, since it was managed as an oligarchy, floundered as Henry Ford's mental powers declined.

Sloan's one crucial oversight was failing to envision a larger, richer role for the factory worker, who was for him merely another interchangeable part. After 1937, when the United Automobile Workers Union was formed to oppose poor labor management practices and bargain for better pay, a formal division evolved in factories that lodged all decision-making and problem-solving with managers. Workers were expected to show up on time, work hard, and keep their opinions about how the plant should run to themselves. Periodic experiments, like the "quality circles" of the 1970s, in which workers were invited on an *ad hoc* basis to suggest improvements, were regarded by unions and workers as insincere.

Toyota learned that to transform mass manufacturing into a highly efficient enterprise took dedication and years of effort. Almost twenty-five years passed from the decision in the 1950s to involve the workers until Toyota was able to adapt quickly to changing market conditions overseas. Gains came slowly and in small increments. Patience was required to see the payoff from seemingly tiny improvements. To make sure that workers remained willing to suggest improvements, the automaker had to be willing to try many suggestions that didn't work.

During the period after World War II, when many Japanese were looking for work, Toyota decided it had to be selective about who it hired in its factories. Based in Nagoya, Toyota chose its workers from the farming communities that surrounded its manufacturing complex. The factory floor demanded more than a mere willingness to show up on time and work hard. Jobs were not only physically taxing but required commitment to *kaizen* and sound basic education in order to keep up with the constant barrage of charts, graphs, and bulletins measuring defect rates and production speed. The job also required better than average interpersonal and communication skills to facilitate harmony in group problem-solving discussions.

Finding the right people for Toyota's factories in postwar Japan wasn't as easy as one might imagine. By the time of Sloan's retirement in 1955, U.S. automakers were pursuing a "move the metal" strategy to meet robust demand. Anyone who could stand on two feet could join the UAW. Toyota, meanwhile, had to experiment with methods to conserve resources as demand for cars in Japan ebbed and flowed. Japanese labor unions, created by General Douglas MacArthur, had won job protection for autoworkers in the early 1950s. Since autoworkers were promised virtual lifetime employment, Japanese automakers lacked the flexibility of American automakers to hire or lay off workers in accordance with the market's demand. So Toyota determined that workers, as permanent assets of the factory, must contribute more than manual labor. Ohno introduced *andon* cords at every work station that could be pulled to stop the assembly line when a worker felt a problem needed to be fixed. Workers were encouraged to pull the cord, repair the problem, and propose solutions to prevent their recurrence.

Toyota's protocol for problem-solving was exceedingly thorough. Workers were encouraged to trace every problem beyond its apparent cause to its root cause. The search for a root cause might unfold in the following way: a worker notices a scratched fender. The fender is repaired; in a meeting of co-workers the worker raises the question of why the fender was scratched. An investigation reveals that the scratch was caused by a misadjusted machine. But why was the machine out of adjustment? Because a worker failed to adjust it. Why did that happen? Because the worker wasn't trained properly. Why was training inadequate? And so on, until the root cause of the problem was determined. Toyota called the process "the five whys."

Though time-consuming and mentally demanding, "the five whys" helped Toyota instill manufacturing discipline. The endless quest for solutions and root causes broadcast the powerful message throughout the company that it was serious about quality and willing to tackle problems head on. The role of the individual worker, therefore, was critical and highly respected by managers. Moreover, workers received the sincere message from their bosses that quality took precedence over production goals. An American could easily imagine a nightmare of workers pulling *andon* cords every five minutes, constantly bringing the factory to a screeching halt. But Ohno discovered that after many stoppages, production efficiency eventually improved as problems were solved thoroughly and workers began to find ways to prevent them.

Another discipline used at Toyota was "just-in-time inventory," or *kanban*. Instead of accumulating large stacks of parts to make sure the assembly line never ran short, Toyota made parts or accepted them from suppliers only as fast they were needed for assembly operations. Consequently, Toyota and suppliers needed to invest less capital than competitors in material and unused parts. But *kanban* made the production system fragile. If a supplier failed to deliver a single part on time, the assembly line stopped. There were no spares. Suppliers and Toyota operated with an understanding of mutual dependence, closeness, and trust. To reinforce this relationship, Toyota bought equity stakes in many of its suppliers, bankers, shipping companies, trading companies, insurance companies, and vice versa, thus creating a family of interlocking relationships known in Japanese as *keiretsu.*

Toyota suppliers quickly learned that the same manufacturing excellence was demanded at parts factories as at Toyota final assembly plants. Instead of buying dedicated machines and producing large numbers of the same part, suppliers learned to operate flexible machines. Though worker training was more extensive, the outlay for machines was reduced, which meant a savings in investment capital. Moreover, if a misaligned machine produced a bad batch of parts, the amount of scrappage was minimized.

In recent years, *keiretsu* has been painted by Western critics as a sinister, exclusionary organization aimed at keeping out Western business interests. Nothing could be further from the truth. Toyota pursued close relationships with its suppliers because it knew that *kaizen* (continuous improvement) flowed inevitably from trusting, open relations with an affiliate. Toyota pushed openness a step further, encouraging suppliers to trade insights among themselves about doing business with Toyota. One by one, automakers enlarged their own *keiretsu,* though that didn't preclude a supplier from selling to other automakers.

Keiretsu proved invaluable for companies in trouble. When Mazda floundered in the mid-1970s due to a series of marketing and technical errors, Sumitomo Bank, a member of Mazda's horizontal *keiretsu* (as distinct from the vertical *keiretsu* of parts suppliers), stepped in and sent executives to Mazda to untangle its problems and provide financial expertise. On countless other occasions, automakers helped *keiretsu* suppliers that were in trouble. As MIT researchers astutely pointed out, a similar system of cross-ownership might have proven invaluable for Chrysler in 1979. Rather than crawl to the government on its knees, Chrysler might have asked affiliate bankers and suppliers for short-term assistance. Members of a Chrysler *keiretsu*

THIRTY-FIVE CENTS

APRIL 17, 1964

TIME

TIME

THE WEEKLY NEWSMAGAZINE

Bertrand son

FORD'S
LEE IACOCCA

VOL. 83 NO. 16
(REG. U.S. PAT. OFF.)

As vice president and general manager of Ford Motor Company's Ford division, Lee Iacocca had the position and power to claim credit for the spectacularly popular Ford Mustang. His energetic PR man managed to get him on the covers of *Newsweek* and *Time* in the same week.

In 1977 Lee Iacocca *(left)* realized that his days at Ford were numbered when Henry Ford II *(right)* announced that Iacocca would share power with Philip Caldwell.

Iacocca's celebrity, and lots of practice, made him a highly credible pitchman for Chrysler's vehicles. But he could not convince most buyers that Chryslers were as well built as Hondas, as he tried to do in a 1990 TV commercial.

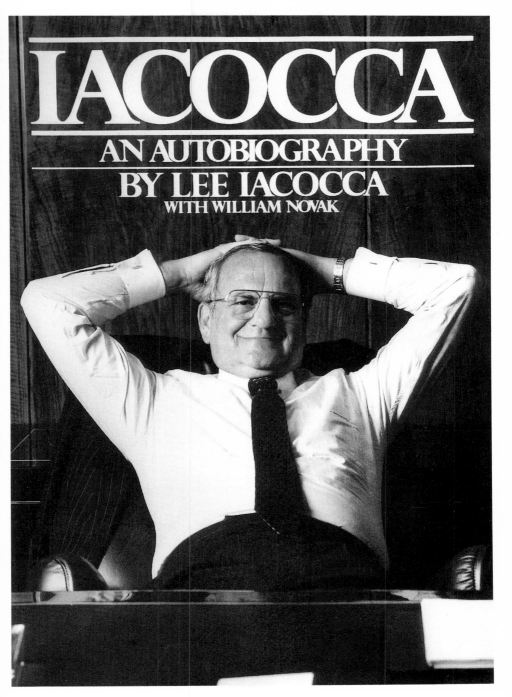

IACOCCA

AN AUTOBIOGRAPHY
BY LEE IACOCCA
WITH WILLIAM NOVAK

Iacocca's 1984 autobiography was a huge best-seller. Its version of how Iacocca overcame unfair persecution and firing by his boss, Henry Ford II, may have been debatable, but it appealed to millions of ordinary working people.

In 1957 the Toyota Motor Corporation exported its first car model, Crown, to the United States. The Toyota Crown was hopelessly underpowered for the U.S. market. Toyota immediately returned to the drawing board to create lighter, stronger, nimbler cars that American customers found pleasing. By the 1970s Toyota was a major importer and a respected force in automaking.

Iacocca battled with the Reagan administration in the 1980s. He blamed Reagan for failing to curb Japanese auto imports and was angry when Treasury officials refused to forego profitable stock warrants they had acquired in return for guaranteeing $1.5 billion of Chrysler loans.

PHOTO COURTESY OF ROBERT LUTZ

DETROIT FREE PRESS

(Above) Henry Ford II snatched Bob Lutz *(right)*, a brash, cigar-smoking American auto executive, from BMW in 1974 and by 1979 had placed him in charge of all Ford's European operations.

(Left) Hal Sperlich was instrumental to the development of the Ford Mustang and later created vehicles for Iacocca at Chrysler. As president of Chrysler he begged for more capital to develop vehicles and was known for his outspokeness. He was finally pushed aside by Iacocca in 1988 to make way for Bob Lutz.

Ben Bidwell, Iacocca's trusted lieutenant, decided to take early retirement in 1990, as Chrysler's financial woes were mounting.

Gerald Greenwald served as number two under Iacocca for more than a decade. Several times Iacocca told Greenwald that he would be his successor. It never happened. Frustrated, Greenwald resigned from Chrysler in 1990 to lead an abortive employee buyout of United Air Lines. Four years later a second buyout succeeded and Greenwald became United's chairman.

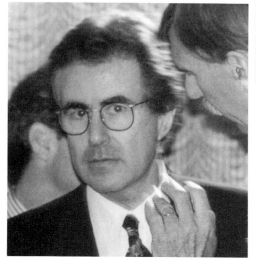

PHOTO COURTESY OF ROBERT S. MILLER, JR.

Robert S. Miller, Jr. *(center)*, known as Steve, was a key player during negotiations with banks and the government in the early 1980s. Miller, a Harvard-trained lawyer and finance expert recruited from Ford, supervised the expansion of Chrysler's financial services subsidiary. He hoped to succeed Iacocca but ultimately Iacocca drove Miller out of Chrysler.

DETROIT FREE PRESS

After numerous clashes, Iacocca *(left)* was determined to prevent Bob Lutz *(right)* from succeeding him as Chrysler's chairman. In early 1992 Iacocca recruited Bob Eaton *(center)*, who had been running General Motors's European operations. Lutz, though disappointed, decided to stay on as number two to Eaton.

Iacocca, with President Bush and other automotive leaders, accomplished very little in a business–government mission to Japan to reduce the bilateral trade deficit. In an address before the Detroit Economic Club in January, 1992, Iacocca complained that the Japanese "are beating our brains in."

Kirk Kerkorian, a billionaire investor from California who knew nothing about the automobile industry, bought a sizable stake in Chrysler in late 1990. In early 1994 Kerkorian launched a hostile bid to buy all of Chrysler. Kerkorian was joined in his efforts by Iacocca, who had retired. Iacocca probably hoped the buyout would increase his already considerable wealth; and he probably yearned for the celebrity and the corporate power that had largely evaporated after his retirement from Chrysler in 1992.

might have quietly resuscitated the automaker without a *de facto* bankruptcy proceeding of the sort held by the U.S. Treasury in administering the Loan Guarantee Act.

The atmosphere of trust promoted by *keiretsu* brought many unanticipated advantages. Toyota expected *keiretsu* suppliers to help design car parts; thus, the automaker was able to entrust new-vehicle plans to suppliers much sooner than Western automakers could. With more time, parts designers were able to suggest numerous innovations and improvements in design. Western automakers, by contrast, were particularly cautious about entrusting suppliers with their designs. They feared espionage from rivals and so designed most parts themselves and then distributed the drawings to suppliers as close to the start of production as they dared.

Western automakers had another motive for controlling the design of their parts: to drive the price of parts as low as possible. By controlling the design, the automaker could terminate suppliers that produced low-quality parts or slipped on delivery schedules by simply handing the blueprints to another supplier. But the bidding procedure often left suppliers with little time to solve basic manufacturing and design glitches.

Western parts suppliers typically bid as low as possible to win contracts. And the winning bidder would often manufacture parts at a loss for some time, in the hope that it could recover costs later with a price increase. Suppliers adopted this tactic partly in self-defense because automakers often switched contracts among suppliers to keep prices low. The cat-and-mouse bargaining rituals between automakers and suppliers were conducted in an atmosphere of mutual wariness. But the process was seldom hostile since it suited both sides. Each knew what to expect from the other. Delays and cost overruns, while regrettable, were absorbed eventually by the buyers of cars as higher prices, or by shareholders as losses.

If anything, automakers had grown quite cozy with suppliers. Longtime suppliers to the auto industry became exceedingly prosperous. The mansions of the families that made hoses and belts and seatbelts and tail lights dotted Bloomfield Hills and Grosse Pointe;

their summer homes lined the shores of Lake Michigan near Harbor Springs and Petoskey; annual profits from supplier companies nourished the boutiques of Birmingham and the glitzy shopping malls north of Detroit. Now and then a minor scandal arose when a purchasing manager from an automaker was discovered to be accepting gifts or cash from a supplier. Officially, automakers adhered to strict policies against gratuities from suppliers. Payoffs were clearly unethical, if not illegal; and of course they added unnecessary cost to a car. Bottles of liquor or fruit baskets at Christmas officially were frowned upon. Unofficially, however, commerce between automakers and suppliers routinely and historically was carried out over sumptuous meals at expense-account mills, during rounds of golf, or at posh hunting lodges, beach houses, and ski resorts maintained by parts manufacturers for entertaining auto executives.

In Japan, the American psychology of parts pricing was turned on its head. Planners meticulously calculated in advance the expected market price of an entire new vehicle, and, therefore, the target price of every part, down to the last bolt. Japanese automakers decided what they were prepared to pay suppliers for each part, not the other way around. Therefore, the burden was on the suppliers to drive their costs lower so as to be able to turn a profit. Suppliers were forced to be efficient, since it was impossible to shift cost to the automaker; and they had to trust the automaker to calculate a reasonable price. Disciplined pricing left little margin for error and encouraged automakers and suppliers to cooperate rather than try to bamboozle one another with crafty bidding tactics. In at least one respect the Japanese did resemble their American counterparts: they too loved mixing business with pleasure, though not on the same scale as in America. A favorite pastime of Japanese automakers and suppliers was a long drunken evening in the "hostess bars" of Tokyo's Ginza neighborhood.

The way Toyota conceived, designed, and engineered new models was swift, efficient, and relatively inexpensive—in short, utterly unlike the unwieldy method of product development that had been institutionalized by American automakers. In America, designers drew pictures and built clay models based on orders from executives and planners. Once the basic exterior design and dimensions were set, the drawings were "thrown over the wall," in Detroit's parlance, to product engineers, who created blueprints for every one of the vehicle's 15,000 or so parts. The product engineers then threw blueprints "over the wall" to manufacturing engineers. The manufacturing engineers, who were entitled to say next to nothing about the design of the vehicle, now had to figure out how to build it. They configured the assembly line and ordered tools, sometimes with great difficulty since the designs could be simple to draw but very difficult to build.

Developing a particular car sequentially from the conceptual phase to design phase to product engineering to manufacturing engineering took American automakers five years or more and was a process steeped in conflict. Each critical function was carried out by a department with its own leader, budget, and staff. Members of departments were discouraged from talking with one another. According to protocol, the chain of command had to be respected. If subordinates from different departments needed to transact business, the subordinate must speak first to a boss, who alone was permitted to speak to his counterpart in the other department. Underlings from different departments were discouraged from exchanging significant information, since information was power—something to be wielded in interdepartmental conflicts. The informal rules of comportment gave rank-and-file engineers anything but a broad view of their mission. Engineers were often closeted in narrow specialties for years and sometimes decades, never gaining an appreciation for concepts such as design for easy manufacture.

Rivalry among departments was fierce. The heads of departments were often jockeying for the same promotion slot. The great dream of every ambitious engineer was to climb the ladder and one day make vice president. To resolve conflicts affecting car designs, however, compromises were vital among ambitious executives. If a vehicle were too heavy, it might mean that an accessory—a rear ashtray, for example—would be targeted for weight reduction. In an atmosphere of cooperation, someone might be willing to try to design a lighter version of a rear ashtray. But in an atmosphere of conflict, the rear ashtray gets canceled and a passenger may jolly well flick his or her ashes out the window. In any event, interdepartmental conflict meant that vehicles rarely emerged as envisioned; often they were so late and over budget that they returned little or no profit on investment.

As Japanese competition made delays and cost overruns a critical issue, American automakers experimented with a watered-down version of team development. A senior engineer or planner was usually assigned interdepartmental authority over groups of engineers and planners. Team members, however, were on loan to the group, straining their primary loyalty to the departmental boss, whose opinion was still most important when bonuses and promotions were handed out. This half-baked system, known as "matrix management," led to great confusion since engineers and planners sometimes did not know who they were working for. Complicating matters, the team leader typically served for only a short period before being reassigned or promoted. A single car project, therefore, might be managed by three or four bosses during its lifetime. Each leader was thus protected from responsibility for failure, though all could claim a measure of credit if the vehicle was a success.

Toyota, by contrast, selected an experienced, charismatic leader to lead a multifunctional group of planners, engineers, manufacturing specialists, marketers, and any other specialist needed to develop a car. The leader was endowed with great decision-making authority. Ultimately, he was responsible for carrying out the vehicle's original mission and for finishing it on time. Inside the company, engineers might say (as they said about the Honda Accord), "That's Miyoshi-san's car." Members of his group were selected and rewarded for cooperativeness and devotion to the ultimate mission of producing a worthy car. (If the car failed, the leader accepted responsibility and might never lead a car project again.) Product engineers worked side-by-side with manufacturing engineers and were expected to consider factors like ease of manufacture and the role of the assembly-line worker when drawing parts blueprints, since ease of assembly resulted in higher quality, which ultimately served customers. To help engineers and planners broaden their view, assignments were varied. Manufacturing engineers routinely were assigned to product design; product designers spent time working in factories.

Conflicts within Japanese vehicle groups arose, but they were resolved promptly by the *shusa*, the boss who bore the ultimate responsibility. Currying favor to enhance one's pay or to win a promotion in Japanese companies was rare since promotions were based mostly on seniority. Car development groups in Japan operated efficiently for at least one reason in addition to their highly rational structure: the Japanese are acculturated from childhood to put the will of the group, classroom, or family above their own personal aspirations. This characteristic of Japanese culture blended neatly with a key teaching of Deming, who was revered within the Japanese business world. Deming taught Japanese managers that perks and pay should seldom, if ever, be used as a performance incentive because they create jealousy and rivalry rather than cooperation among co-workers.

Kim B. Clark and Takahiro Fujimoto, who had studied American and Japanese styles of product development while at the Harvard Business School, provided the International Motor Vehicle Program with striking empirical evidence of the differences between the two systems. In the mid-1980s, according to their research (published in 1991 as *Product Development Performance*), the Japanese auto industry needed, on average, 1.7 million engineering hours to develop a new car, compared with 3.1 million hours for the U.S. auto industry. The average development cycle took 46.2 months in Japan, compared with 60.4 months in the United States. The average number of employees assigned to each project was 485 in Japan, compared with 903 in the United States. Moreover, the American auto companies were significantly slower in bringing a new car design into production; they were slower to reach full production speed and slower to achieve a normal level of quality.

Any automaker able to minimize the cost and lead time of new products gained obvious advantages over slower rivals by reaching the marketplace first and capturing the attention of the customers. Honda, by swiftly and cost-effectively creating ever more sophisticated versions of its Accord sedan, produced the best-selling car in the United States in 1989, the first time a non-Ford or non-Chevy model had achieved that distinction in eighty years. Honda decided to introduce an all-new Accord every four years while the Big Three were standing pat for five to ten years with basic midsize car designs.

Changing models quickly was impossible without an agile workforce and lean, flexible factories. Year after year, Toyota factory workers were encouraged to suggest manufacturing improvements and ways to better perform their duties. The measure of Toyota's manufacturing skills was literally unbelievable to many American executives when they first perused the MIT data gathered by Womack and his associates. At GM's Framingham plant—a facility typical of GM, Ford, and Chrysler's older assembly plants—thirty-one hours of labor were needed to assemble each car, compared with sixteen hours per car at Toyota's Takaoka plant. Cars built at Framingham registered an average of 131 defects per car versus forty-five defects at Takaoka. The Framingham plant could operate flat-out for two weeks without depleting the parts kept in inventory; at Takaoka the inventory was so spare that the assembly line would run out of parts in two hours without replenishment. Obviously, Toyota was investing a fraction of what General Motors was investing in parts inventory.

The Machine That Changed the World didn't glorify everything Japanese; nor did it fail to recognize that the American auto industry was awakening to lean manufacturing. Nor did every Japanese automaker achieve Toyota's productivity or quality. Some Ford plants, in fact, had narrowed the gap with Japanese counterparts. Ford, which owned a 25 percent stake in Mazda, launched a companywide drive in the early 1980s, called "employee involvement," to learn from Japanese automakers and to encourage suggestions and ideas from Ford workers. Many Ford managers who had relied on intimidation to motivate subordinates and weren't willing to change were demoted or retired. Moreover, Ford developed its Ford Taurus/Mercury Sable midsize car using a modified version of Japanese product development. Having earned a reputation for sturdiness and quality, the Taurus rose steadily in popularity from 1985 to 1992, finally unseating the Honda Accord as the nation's best-selling car.

While American automakers were at least demonstrating a tentative curiosity about lean manufacturing principles, the Europeans appeared hopelessly backward in the late 1980s. As in America, Japanese automakers had opened plants in Great Britain and wanted to do so on the continent. France, Great Britain, Spain, and Italy erected formal import barriers against Japanese cars in order to protect their native auto industries. In 1992 the European Economic Community decided to limit Japan's market share on the entire continent, the principal result of which was to push lean manufacturing's impact further into the future and prolong the inefficiency of European car plants. To Europeans, the reduction of automotive jobs that accompanied lean manufacturing was politically unacceptable. Such automakers as Peugeot, Rover, and Fiat—and to a lesser extent, Volkswagen and Mercedes—were permitted, by virtue of chauvinistic trade politics, to hang on a bit longer than U.S. counterparts to outmoded mass manufacturing. In a few countries, like Belgium, where consumers were freer to buy imports, the Japanese automakers were making great inroads in the market, a portent for what lay on Europe's horizon.

Europe was even less eager than America to consider the breathtaking consequences of mass manufacturing's demise. Social wreckage in the form of closed plants, lost jobs, and shattered communities was too painful and too politically risky to contemplate. To Womack and his associates though, the arguments for swift introduction of lean manufacturing—besides its inevitability—were overwhelming: lean manufacturing provided satisfying, fulfilling jobs. The methods of lean manufacturing enhanced competitiveness without enormous capital outlays.

Unless Detroit learned lean manufacturing, the Big Three were bound to face more vicious competition in their own backyard from plants like the one owned by Honda in Marysville, Ohio—just a three-hour drive down I-75. American automakers had to learn how plants like Honda's really worked, and quickly.

Unless Detroit swallowed its pride and studied at the knee of the Toyotas and Hondas of the world, its survival was in question.

ASIA IN AMERICA

ON A SUMMER EVENING in 1982, in a striptease club outside Detroit, a fistfight erupted between Vincent Chin, a twenty-seven-year-old Chinese-American who was being fêted at a bachelor party, and a group of strangers who had been taunting Chin and his friends.

The scuffle might have ended unremarkably, like most barroom brawls, but a short time later Ronald Ebens and his stepson, Michael Nitz, two of the men who had been baiting the celebrants, spotted Chin in front of a McDonald's in Highland Park, a small municipality bordering Detroit. Ebens, an unemployed autoworker who had last worked in a Chrysler plant, grabbed a baseball bat from the trunk of his car, chased Chin a short distance, and, in a drunken fury, clubbed the man to death while Nitz held him down. A short time later the two men were arrested. After pleading guilty and no contest, respectively, to charges of manslaughter, Ebens and Nitz were placed on probation and fined $3,700 each.

The victim's family and numerous Asian-American civil rights groups expressed outrage that the sentences were not heavier. They maintained that Chin had been killed for one reason, because he was Asian. During trials and hearings held in the wake of his death, witnesses testified that before the fight someone had called Chin a "nip" and other derogatory racial terms. Another witness said she heard someone say, "Because of you motherfuckers we're out of

170

work." The defendants evidently believed Vincent Chin was Japanese or Japanese-American.

Defense attorneys argued that the attack was not racially motivated, that it was a simple barroom fight that turned deadly. Chin, they noted, had thrown the first punch.

A federal grand jury suspected racial motivation for the crime and indicted Ebens and Nitz on charges of violating Chin's civil rights. In 1984, Ebens was convicted. Three years later a federal appeals court overturned the conviction, ruling that the judge in the first civil rights trial wrongly disqualified evidence suggesting that witnesses to the race-baiting had been coached. Ebens subsequently was retried and acquitted.

Notwithstanding the complex circumstances of the case, Asian-American and civil rights groups regarded Vincent Chin's death as a violent and frightening watershed of the racial hatred against Asians that had grown out of the climate of economic friction between the United States and Japan. Countless incidents, some quite ugly, had contributed to the worrisome trend of hostility directed at Japanese and other Asian minorities. In fact, the trend had a long history. Racial discrimination, aggression, and misunderstanding have characterized relations between Japan and the United States for nearly the entire 140 years since Japan ended its self-imposed isolation. After World War II, when anti-Japan sentiment reached a peak in the United States, hostility against Asian-Americans remained intense but didn't receive much attention compared with black-white racial tension. Since the early 1980s, a time when the U.S. auto industry was depressed and thousands of workers were wondering if they would ever draw a paycheck again, civil rights experts began to note an upswing in violence and discrimination against Asian-Americans.

Every problem needs a cause, and every outrage needs a perpetrator. In the matter of automotive factory shutdowns and the loss of hundreds of thousands of jobs, leaders of the U.S. auto industry were quick to accuse Japan and its auto industry of targeting the American automobile business for conquest. They also blamed U.S. government officials for failing to protect the American market. Now and then, the automakers actually blamed American consumers as well, for choosing Japanese brands and abandoning their loyalty to the cars they had always bought.

Workers and communities wounded by the automotive downturn demanded to know the cause of their suffering. Politicians from Michigan and other automotive centers in the Midwest sometimes championed the auto industry's vilification of Japan in blatantly racist terms. John Dingell, Democratic chairman of the House Energy and Commerce Committee, who represented a suburb of Detroit, once referred to the Japanese auto industry — in an open session of Congress — as "those little yellow people." In suburbs of Detroit, Japanese cars and trucks, including vehicles built at Japanese plants in this country, became common targets for vandalism, particularly in neighborhoods where many autoworkers lived. Union officials encouraged aggressive attitudes by posting hostile messages warning drivers of foreign cars not to park near factories or union offices. Workers, their families, and neighbors displayed bumper stickers with ominous messages like "Toyota. Honda. Pearl Harbor."

Now and then a union local or car dealer sponsored a "bashing" event, which consisted of towing a Japanese-built car or truck to a parking lot and inviting frustrated workers to pummel it with a sledge hammer. The atmosphere of aggression set the stage for acts of violence that were to follow. The American auto industry, as workers perceived it, was struggling for its life. Factory workers had been accustomed to periodic layoffs and callbacks. No one had worried too much when the economy softened and the assembly line slowed. Unemployment checks helped families get by; some laid-off workers found odd jobs and others enjoyed the time off. But when permanent plant closings hit the steel and auto industries, panic set in. Factory gates were closed forever. Weeds pushed through the parking lot asphalt. Gone were the middle-class paychecks that had supported workers' families. Gone were the tax revenues that paved roads, bought textbooks for school children, and sustained social welfare agencies.

Blue-collar hostility to Japan took on a rough edge, translating into tire slashings, shouted curses, and the occasional Toyota reduced to a battered hulk. White-collar hostility to Japan showed a more genteel face. Auto executives, worried by competition from Japanese manufactured goods, lashed out in speeches and lobbied government officials, citing the trade deficit with Japan as proof that American economic well-being had been seriously compromised.

The U.S. auto industry's fixation on trade-deficit numbers wasn't surprising since they were caused mainly by the flow of cars and automotive parts from Japan. Until 1976, trade between Japan and the United States was roughly in balance. In other words, Japan imported from the United States roughly as much, in value, as the United States imported from Japan. From then on, imports from both countries grew, but United States imports from Japan — particularly automotive imports — grew at a much faster rate. Then suddenly in the 1980s the value of Japan's imports, mostly agricultural commodities, oil, and other natural resources, collapsed in the worldwide deflation of commodity prices. Economists debated the deficit's importance, but U.S. business viewed it with alarm. In 1987, the overall U.S. trade deficit stood at $152 billion, $56 billion of which was the deficit with Japan. Of the $56 billion deficit with Japan, $33 billion reflected the flow of vehicles and car parts to the United States.

The flurry of raw trade-deficit numbers diverted attention from significant trends. Japan's total trade surplus was much larger than its specific surplus with the United States. And the U.S. deficit worldwide was much larger than its deficit with Japan alone. Japan's surplus and the U.S. deficit had more to do with particular characteristics of the respective economies of the two countries than with the trade relations between them. Japan had concentrated on manufacturing and done well at it. American consumers, the most prosperous in the world, were the prime customers for Japanese goods. American manufacturing, for many reasons, didn't keep pace with improving Japanese goods, to the detriment of U.S. exports. Nevertheless, critics who focused on the negative U.S. trade statistics with Japan tended to overlook the fact that the United States sold more goods and services to Japan than to any other trading partner except Canada.

Executives and politicians, speaking on behalf of American

industries caught in the competitive squeeze, warned that Japan was enhancing its national strength by classic mercantilistic tactics. Their implication was that Japan was suctioning America's wealth. By exporting more than it imported, Japan was able to amass a substantial financial surplus. But instead of stockpiling gold, as monarchs of old, Japan was spending its surplus abroad, buying assets such as office buildings in Los Angeles, the Columbia Pictures movie studios, and U.S. Treasury bonds.

Japanese investment in U.S. assets and in U.S. government debt provoked worry. These worries, though, were vague and unformed. No one could really explain why a Japanese company owning Rockefeller Center should be seen as sinister, or why—aside from blatant racism—ownership of American assets by English or German companies wasn't equally scary. On this point the less educated view was the more honest. Those who were suspicious of "the sneaky Japanese," still harkening back to the surprise attack on Pearl Harbor, were convinced that the Japanese who owned U.S. assets had to be up to something nefarious.

Many economists held a less panicked view of trade. relations between the two nations. Despite differences in culture, Japan and the United States shared basic values when it came to industrial competition. As economic powerhouses, they depended on one another. Japan's automakers flourished as a result of their ingenuity, talent, and determination—and what they learned from the United States. Secondarily, Japanese industries' chief rivals—the U.S. steel industry, the U.S. electronics industry, and the Big Three—had simply fallen short in the areas of cost, quality, and service. As buyers increasingly chose cars imported from Japan, trade-deficit figures inevitably reflected those choices.

Economists didn't necessarily regard a large bilateral trade deficit as reason for alarm. The United States was the undisputed world leader in many industries—computers, pharmaceuticals, aircraft, to name only a few. Though persistent bilateral deficits with Japan weren't healthy, they were less crucial than the American trade imbalance worldwide. One key to reducing the trade deficit was exporting more, a policy that was doomed to fail as long as the United

States intimidated important trading partners like Japan in order to satisfy certain parochial interests.

Plenty of American companies, particularly those that exported heavily, opposed trade barriers. As history had demonstrated time and again, trade barriers inevitably provoked retaliation. (The Smoot-Hawley Tariff Act, passed in the wake of the 1929 stock-market crash, was regarded by many economists as the single most important cause of the Great Depression.) Reduced commerce between nations led inevitably to less prosperity for everyone. America's balance of trade was in deficit, but the nation exported huge amounts of goods and services and couldn't afford to be shut out of overseas markets.

Detroit, of course, viewed the trade deficit more narrowly. Japan's "voluntary" export restraints on cars had backfired. Japanese automakers had simply shifted exports from small cars to more expensive, higher-profit models—Lexus, Acura, and Infiniti—while accelerating factory construction in the United States. Overnight the U.S. industry was contending with Japanese competition in the lucrative luxury segment of the market, once dominated by Cadillac and Lincoln. Japanese "transplant" factories were soon building all sorts of models. Detroit called for barriers more effective than voluntary restraints had been; and to sell such barriers the industry needed a spokesman who could make people understand and believe the nature of the Japanese threat.

Lee Iacocca—shrewd, loquacious, and credible—was a natural for the job. For Iacocca, demonizing Japan served two purposes. First, he hoped to win over politicians who remained skeptical of protectionist fixes for the auto industry's problems. And, second, his rhetoric was designed to convince consumers that buying Japanese cars didn't just hurt companies like Chrysler but undermined U.S. economic strength in general. In a coordinated public relations effort, Iacocca meshed the imagery and ideology of Chrysler advertising together with his speeches and fund-raising on behalf of the Statue of Liberty celebration. When General Motors announced its Saturn small-car project, Chrysler instantly countered with a project called "Liberty." Chrysler advertising employed patriotic theme songs; the automaker went so far as to name one of its models "America."

With Chrysler advertising as his backdrop, Iacocca preached that American-built cars were more than smart buys, they were counterpunches against Japan. Iacocca's speeches also attacked American trade policy, which had done little to ease the trade deficit. In 1986, Iacocca proposed that the United States impose on Japan a target of reducing our trade deficit with them by 20 percent per year for five years. How Japan was to reach this target was its own concern. "We don't have to be bullies about it," he wrote in a syndicated newspaper article. Liberal U.S. trade policy had benefited Japan, but enough was enough. "We've earned the right to make some firm demands."

The Republican administration of Ronald Reagan, with its ideological commitment to free trade, had little choice but to show it was willing to act to help balance the trade deficit. Under the terms of the Plaza Accord of 1985, Japan agreed to cooperate with monetary strategies that would raise the value of the yen and lower the value of the dollar. The currency swing made American exports less expensive and Japanese imports more expensive. If Detroit was serious about redressing the trade balance, in the view of the administration, it had to sharpen its competitive skills and export more. (Detroit automakers didn't help their case by aggressively importing Japanese machinery, engines, and fully built vehicles. Chrysler imported hundreds of thousands of Mitsubishi engines and Mitsubishi vehicles, which it sold under the Chrysler name.)

Despite his efforts, Iacocca learned that tough protectionist legislation was a nonstarter in Washington. In the view of Republicans and many Democrats, protection of a vital industry didn't justify a trade war. "A politician would rather be called a pervert than a protectionist," Iacocca wrote in a bitter syndicated newspaper column in 1987. Increasingly, Iacocca found himself on the defensive when confronted by the press. Why, reporters hectored him, should American consumers be prevented from buying Japanese cars when Chrysler regularly shopped for engines, parts, and vehicles in Japan?

As Iacocca grew frustrated by the lack of political response, his speeches and writings took on a menacing tone. In 1987 he warned that Japan's "attitude has already convinced many Americans that the

Japanese have to be forced to change because they'll never do it on their own." If the Tokyo government fails to address trade deficits, he said, "then sooner or later the American government will have to, and the storm might turn into a typhoon."

Two years later, Iacocca joined the chorus of voices — inspired by Pat Choate's book *Agents of Influence* — attacking Japan's lobbying in Washington as a clandestine effort to manipulate U.S. economic policy. Lobbyists for Japan, Iacocca wrote, "get the American government to open its kimono when they want a little more of this market or want to protect what they already have."

With more plants in the United States and growing U.S. sales, Japanese companies indeed had hired more lawyers and lobbyists to represent their legitimate interests. Japanese lobbyists successfully argued that the 25 percent tariff on imported light trucks should not apply to minivans or other models that primarily carry passengers. But why was Japan being singled out? Washington was influenced by special interests of all kinds, foreign and domestic. Lobbyists working for Japanese companies had obeyed the law. No one had ever questioned the right of English, Dutch, or German companies to lobby here.

The Japanese lobbying that provoked Iacocca's indignation recalled Detroit's own failed efforts to influence policy and legislation. In its eighty-year history, the U.S. auto industry's lobbying efforts in Washington were weirdly ineffectual and awkward compared with those of much smaller U.S. industries. This was so, in part, because the Big Three could agree on very little among themselves and were constantly attempting to gain an advantage over one another. Ford and General Motors had shown little sympathy for the government's bailout of Chrysler. Chrysler tried to block GM's joint venture with Toyota on antitrust grounds. The three had clearly differing positions on energy conservation and gasoline taxation. They differed regularly on tactics to oppose safety regulations and air-pollution rules.

Time and again, members of Michigan's Congressional delegation urged U.S. auto executives to concentrate their potentially enormous influence by reaching a consensus and then speaking with one voice.

Lawmakers from Michigan and other auto-sensitive regions weren't able to horse-trade effectively when their fellow legislators were confused about what the industry really wanted. As one frustrated representative liked to quip, "Three minus one equals zero."

(In 1989 chief executives of the Big Three held an extraordinary meeting at Ford headquarters to discuss cooperation on many issues, including a united front in Washington. The meeting constituted a tacit recognition that the Japanese auto industry posed a common threat of overriding importance to the Big Three. Within three years, the industry replaced its sleepy trade organization, the Motor Vehicles Manufacturing Association, with a better-financed, more powerful organization, the American Automobile Manufacturers Association.)

Because the automakers weren't strong enough to make protectionism fly in Washington under its own name, Lee Iacocca tried to find a way to repackage it. Master marketer that he was, he redesigned the vocabulary and refocused the trade deficit rhetoric to better express who was to blame and how the problem could be fixed. Iacocca set sail on a fresh course, away from the fight between "free trade" and "protectionism." Instead, he delivered sermons about the need for "fair trade." Denying he ever favored protectionism, Iacocca said he simply wanted fair play—fair trade versus unfair trade. In other words, if Japan was to be permitted to sell cars relatively unfettered in the American market, the Big Three should be allowed to do the same in Japan. But because Japan obviously had closed its home market and prevented U.S. automakers from selling cars in Japan, or so his argument went, we had every right to keep Japanese goods out of the United States.

Of course, Iacocca's stratagem rested squarely on the assumption that U.S. automakers had been trying hard to sell in Japan but were unfairly barred from that market. Statistics showed clearly enough that U.S. carmakers weren't selling many cars to Japanese customers. The reasons for the weak sales, however, were a matter of some dispute. Automotive experts in Japan asserted that most Big Three models were ill-suited for Japan. Few Big Three models were available with the steering wheel on the right, the Japanese standard. Moreover,

American midsize sedans were huge by Japanese standards, far too wide for the narrow parking spaces near Japanese homes.

True, Japan *had* booted American automakers out of the country in the 1950s. But Japan's policies, originally designed to allow the country's infant auto industry to sprout, had since relaxed; yet the American auto industry devoted little time or thought to the Japanese market, beyond scrambling to buy equity stakes in Suzuki and Isuzu (General Motors), Mazda (Ford), and Mitsubishi (Chrysler). With so much sales growth in Europe and the United States in the 1960s and 1970s, who had time to worry about Japan? Consequently, the Big Three automakers had never built a network of retail dealers in Japan, nor had they bothered to study Japanese consumer tastes.

The Japanese were confused and irritated by incessant harangues from America to "do something" about the phenomenal success of their industries, particularly their automakers. The two countries had quarreled for years over what constituted "fair" rules for trade and competition between industries, without reaching a comprehensive understanding. Japan felt it was being punished for its success. Japan's industries felt entitled to be proud of their achievements. That American customers bought Japanese cars and trucks so enthusiastically justified the years of sacrifice and billions in investment to improve early models that had been so embarrassingly second-rate. Surely if U.S. automakers had been building better vehicles, Americans would never have looked at Japanese models. If Americans truly believed that they were threatened by Japanese cars, they were free to buy cars built in Detroit. Such was the principle of free markets.

Censure from American politicians and corporate leaders wounded Japan. The Japanese had been striving to understand Western dress, manners, and business customs since shortly after the day in 1853 when Commodore Matthew Perry steamed into Tokyo Bay to open the country forcibly to the outside world. Japan tried to learn the West's codes of behavior, only to find out that America—at least America as represented by Lee Iacocca and those who supported him—regarded Japanese achievements as threatening.

Within fifty years of joining the international community, Japan had vaulted to the status of world power. Japanese academics and businesspeople had traveled the world, studying Great Britain's navy, Germany's army, and American diplomacy, which formed the country's notions about foreign policy and empire. Foreign experts by the thousands were invited to Japan.

But the European-American world was slow to accept non-whites as equals. History demonstrates that the first commercial treaties negotiated by Western powers with Japan were almost always "unequal," that is, tilted in favor of the Western powers. For example, most Western treaties with Japan gave foreign powers extraterritorial rights, so that its citizens could live in Japan subject only to the laws of their home countries. The same was not true for Japanese living abroad.

Unequal treaties were revised in the 1890s. But the United States and European powers, wary of a non-white people that could rise so quickly to world-power status, were reluctant to give Japan its due in international accords following Japan's military victories over China and Russia at the turn of the century. Japan was acting in an imperialistic fashion, but no more so than other great powers concerned about securing access to natural resources. Each time Japan attempted to expand its influence in Asia, its efforts were met with Western resistance and a sense of outrage that Japan would dare mimic the behavior of the white superpowers in places like China, Africa, India, and the Middle East.

In California, competition for land and jobs among Asian immigrants and white pioneers at the turn of the century ignited an atmosphere of race hatred, prompting the first laws designed to exclude Asian immigrants as "undesirables." Anti-Asian discrimination was becoming institutionalized: for example, San Francisco segregated its school system to keep Asians out following the earthquake of 1906. These events were watched closely in Japan, where they were regarded as intolerable insults to the emperor and the proud race of Yamato. With stunning naval victories against Russia at Port Arthur and the Strait of Tsushima fresh in their memories, Japanese editorials howled for armed redress. Genuinely worried that war could break out between the two countries, President Theodore Roosevelt ordered the U.S. Navy's "great white fleet" from the Atlantic to the Pacific in 1908 to quiet the hostile tremors — a fact that comes as a surprise to many Americans, who know startlingly little of their country's relations with Japan prior to Pearl Harbor.

World naval conferences of 1922 and 1930 were efforts to bottle rising Japanese power, based on the assumption that Japan had no right to consider itself on an equal footing with Great Britain, the Soviet Union, or the United States. By the 1930s, Japanese militarism and expansion in China, Manchuria, and Southeast Asia were condensing into a fascist brew that exploded with the surprise attack on the U.S. fleet. After its utter destruction in World War II, Japan foreswore militarism. Most Japanese had blindly followed the military clique and the emperor into war, though afterward they were quite willing to admit that challenging the United States had been a worthless idea. Nevertheless, the Japanese people never lost their fierce pride, their sense of indignation, and their deep desire finally to be accepted as equal in the community of great powers.

After the war, Japan was given a fresh start under the supervision of the American occupation led by General Douglas MacArthur. The country immediately resumed its study of the outside world in order to absorb the knowledge it would need to rebuild as a modern industrial society. America helped its former enemy get back on its feet. Our hopes for Japan were mixed with the sudden realization that we needed a counterweight against the two huge Communist societies in Asia, the People's Republic of China and the Soviet Union.

America thoroughly remodeled Japan's economy according to American democratic principles. MacArthur imposed strict antimonopoly laws that called for the dismemberment of the four big *zaibatsu*, or industrial cartels, that were dominant before the war. Later on, the American attitude toward monopolies softened, in part because Japanese firms demonstrated that they were able to strengthen one another technically and financially by cooperating. Under new regulations, the *zaibatsu* were reborn as *keiretsu*, or industrial groups. Unlike the prewar *zaibatsu*, which often were dominated by families

and closely linked with political parties, the *keiretsu* were characterized chiefly by exchange of equity interests among companies, interlocking directorates, and close supply relationships.

The Korean War was a godsend to Japan's economy, creating demand for its goods and stimulating investment in its industries, and
within a few years Japan returned to pre–World War II levels of production. Economic planners exercised strong influence over industry,
which was possible because business leaders felt a strong sense of
duty toward authority and toward the national mission of revitalizing
their country. Due to the scarcity of natural resources and hard currency, industries were directed to concentrate on producing finished
goods for export. Export credits could be used to generate hard currency for importing technology and raw materials. Since the United
States had insisted on the demilitarization of Japan, the country wasn't
obliged to allocate resources on defense. The world luxuriated in the
knowledge that Japan wouldn't be a military threat. The United States,
moreover, was more than happy to provide a big, hungry market for
steel, textiles, and other goods—with very few import restrictions.

Japan only had to peer across the Pacific to see that a healthy
automotive culture was perhaps the best indicator of a prosperous
industrial society. Prior to the 1930s, many people in Japan had never
seen a car or truck. Vehicles in Japan either were imported or assembled locally from kits imported from abroad. Japan's auto industry
was born in the 1930s and was limited, under prodding from the
military, almost exclusively to the production of trucks. Immediately
after World War II, it remained concentrated on trucks, which were
badly needed for construction at home. Widespread poverty limited
domestic car production to just a few hundred per year.

By the beginning of the Cold War austerity in Japan was relaxed,
and the country was promoted to the economic status of the United
States's junior partner. The Marshall Plan was in place to help rebuild
Western Europe, and the United States extended similar economic
recovery programs for Japan. With brighter economic prospects, Japan's automakers proposed a five-year recovery plan with the goal of
producing 120,000 vehicles annually by 1952, mostly trucks. Only
112,000 trucks were in service in the entire country in 1946, many

burning a variety of alternative fuels, including coal. Nissan and Toyota, Japan's two major automakers, were eager to bring motor transportation to ordinary Japanese as well. Their executives had traveled the world and seen the profound impact personal transportation had on every facet of life, from the way people appreciate art to the way families seek recreation.

Automakers envisioned privately owned automobiles swiftly replacing the little three-wheeled vehicles often used for passenger travel. But Japanese automaking know-how, not to mention the country's roads, were woefully inadequate. Japan's government, moreover, was deeply divided over whether to encourage the auto industry to expand from trucks into passenger cars or to direct economic resources elsewhere. One school of thought among bureaucrats doubted that Japanese automakers could ever match the level of their American and European counterparts. The Bank of Japan and the Ministry of Transportation, afraid that the automakers were too weak and unsophisticated, favored leaving cars to the Europeans and Americans and encouraging only local truck production.

But bureaucrats from the Ministry of International Trade and Industry (MITI) prevailed, arguing that local auto production would stimulate the steel, precision machinery, and related industries. Thus carmaking was deemed a "strategic" industry and was charged with helping the nation to achieve self-sufficiency and economic parity with the rest of the industrialized world. Starting in 1949, automakers adopted as their goal to build cars that would sell abroad. To help them reach that goal, the Japanese government required automakers to divest themselves of nonautomaking subsidiaries while permitting them to double their capitalization. Toyota immediately issued bullish projections for an annual production of 21,400 cars and trucks by 1950.

Throughout the 1950s, MITI sponsored the issue of low-interest loans, tax credits, and exemptions from import duties, all designed to help Japanese automakers gain stability and financial strength. The automakers were encouraged to buy as much technology abroad as they could afford. At the same time, the government instituted a series of protectionist regulations designed to minimize competition locally, since Ford and General Motors would easily have put Nissan and Toyota out of business with imports and local production.

Given Japan's lack of automaking experience, its first car models didn't amount to much. Toyota Motor Corporation engineers began work in 1945 on a car engine with one liter of displacement that developed a paltry twenty-seven horsepower. Plagued by all sorts of mechanical defects, Toyota's 995-cc Toyopet fell far short of being a rousing success. Japan's industry needed new vehicle models as well as the capacity to build them. In 1955, MITI, intrigued by the popularity of the Volkswagen Beetle, sponsored a contest among Japan's automakers to create an affordable minicar, a "people's car." MITI wanted each automaker to build to the same specifications: namely, that their minicars have a maximum speed of 62 miles an hour, fuel efficiency of 70 miles to a gallon, and a manufacturing cost of roughly $400.

Japan's automakers favored protection and government credits, but they didn't appreciate heavy-handed direction as to what models they should build. Still, they went along. As minicars like the Subaru 360, Toyota Publica, and Mazda R360 began appearing between 1958 and 1960, the three-wheel vehicles vanished almost entirely. Car production rose from 20,000 in 1955 to 165,000 in 1960 and 696,000 in 1965.

Once Japan began to look more and more like an emerging economic power, the rest of the world expected the country to behave accordingly. Japan joined the General Agreement on Tariffs and Trade (GATT) in 1955, which obliged signatories to liberalize trade and progressively reduce tariffs. Five years after signing, Japan still had eased tariffs on only 40 percent of the items covered under the agreement. Western nations, meanwhile, began applying pressure on Japan—with little effect—to relax restrictions against imports. Japanese bureaucrats were still worried that their infant auto industry might be destroyed if they let foreign automakers operate freely in their domestic market.

MITI tried to mitigate what it saw as destructive competition among Japanese automakers by coaxing them to merge. (By 1961,

Japan was only the seventh largest automaker in the world but had seven automakers, which inhibited them from maximizing economies of scale.) Except for one or two minor mergers, the initiative failed. Feeling more and more independent, automakers didn't appreciate pressure to merge, any more than they did suggestions about which models to build. MITI was ignored when it asked Honda to stay out of the automobile business, another bit of evidence to suggest that the much-vaunted alliance between Japan's government and its auto industry — to be decried years later by Detroit's protectionists in Washington — was not without its drawbacks.

Japan didn't lower the tariffs on cars significantly until 1970, having waited until its industry reached a production level of five million vehicles annually (60 percent cars and 40 percent trucks) and had become the world's most efficient producer. By 1978 tariffs were eliminated completely but any importer who wished to try out the Japanese market had to face a well-entrenched local industry with a sophisticated network of retail dealers. Moreover, nontariff barriers to entry were legion, such as a regulation that required each imported car to be inspected individually, a cumbersome and expensive process seen by foreign automakers as blatant protectionism.

Whether a determined U.S. automaker might have overcome Japan's barriers is moot. Detroit in the 1970s had far more pressing problems, namely compliance with substantive safety, fuel-efficiency, and air-quality regulations for the first time in its history. But the early 1970s would have been the right time to learn about emerging Japanese automotive tastes and consumer habits, to find a right-hand-drive model that appealed to Japanese drivers, and to devise a clever distribution strategy — to put the ball in the opponent's court, so to speak. Instead, the Big Three took advantage of the Japanese relaxation of a ban on foreign investment and bought stakes in local automakers. (General Motors bought 35 percent of Isuzu, Ford bought 24 percent of Mazda, and Chrysler bought 15 percent of Mitsubishi.) In 1970, Japan's car exports topped one million units for the first time, accounting for a fifth of its industry's total production. Acquiring equity stakes, for the moment, was the Big Three's method of gaining a foothold in Japan's automotive sales growth and of keeping closer tabs on the factors influencing the steady increase of exports to the United States.

Between 1970 and 1980, Japanese vehicle exports to the United States surged six-fold, from about 422,000, or 4.2 percent of U.S. sales, to more than 2.4 million, or 21.4 percent of U.S. sales. Almost overnight the U.S. auto industry realized that if it didn't find an answer to the Japanese surge, Chrysler was doomed and Ford wouldn't be far behind. The specter of hundreds of thousands of unemployed UAW members, padlocked factories, and deserted towns made politicians on both sides of the Pacific nervous. The voluntary export restraints agreed upon in 1981, which had been designed to give U.S. automakers a chance to recuperate, spurred strategists from MITI and Japan's Big Six (Toyota, Nissan, Mitsubishi, Honda, Mazda, and Fuji) to devise a longer-term answer to the political dangers occasioned by their success.

The answer was to transplant factories in the United States. This approach neatly resolved several vexing issues simultaneously. Instead of "undercutting" America's automotive employment, the Japanese automakers would be creating jobs for American workers. Instead of adding to the U.S. trade deficit, locally built cars would replace imports that hurt the balance of trade—and they might one day even be exported and actually *help* the U.S. balance of trade. Finally, consumers who had been reluctant to buy imports might begin to think of "Japanese" cars as "American," or at least as having no nationality, like Sony, and to judge them on their merits alone.

Honda, which had already been assembling motorcycles in the United States, was, in 1982, the first Japanese automaker to open a U.S. car assembly plant, in Marysville, Ohio. Honda, at this time still a second-tier player in Japanese automaking, shrewdly determined that it stood a better chance of winning new customers in America than in Japan. Nissan decided shortly thereafter to hire a group of former Ford executives to open an assembly operation in Smyrna, Tennessee, using principles of Nissan's manufacturing doctrine adapted to an American setting. Toyota, the most conservative Japanese automaker, joined with General Motors in 1984 in Fremont, California. (A few years later Toyota built an assembly plant independent of U.S. partners in Georgetown, Kentucky.) Mitsubishi and Chrysler announced their factory for Bloomington, Illinois, in 1985, to be followed by several more Japanese auto plants across North America.

Hoping to avoid the acrimony that characterized relations between the UAW and managements of the Big Three, Honda, Nissan, and, later on, Toyota decided to locate their plants as far away as possible from the union's urban strongholds. The automakers reasoned that the UAW would attempt to organize their workforces and that at some stage their workers were going to vote on whether to allow a union to represent them. When that day came, they hoped workers would feel no need for UAW representation; thus, they wanted to locate in places where prounion attitudes were not traditionally strong.

There was another reason for avoiding large American cities. Aware of the yawning gap in productivity between the U.S. and Japanese auto industries, Japan's automakers wanted a workforce that was young, well-educated, imbued with good work habits, and, most importantly, able to work cooperatively in a group or team. Japanese newspapers wrote constantly about American cities beset with drugs, violence, decay, and alienation. The automakers reasoned that they should scout rural and semirural locations in America's heartland, where job applicants were likely to have been brought up in small communities with solid middle-class values, reared in homes by two parents, and taught in schools that stressed discipline and achievement.

The first American assembly plants operated by Toyota, Mazda, and Mitsubishi were joint ventures with General Motors, Ford, and Chrysler, respectively. Since the Big Three couldn't afford to jeopardize their relations with the UAW by entering into nonunion ventures, they had little choice but to insist that the UAW be recognized as the bargaining agent for workers. However, the Japanese—wishing to institute their own methods and keep labor costs competitive—stood firm in their insistence that the labor agreements not be traditional

UAW contracts. Using its production system from Japan as the model, Toyota hired former General Motors workers and set up operations in a former GM plant in Fremont, California. Similarly, Mazda built its assembly plant on Ford property near Detroit in Flat Rock, Michigan, and Mitsubishi built a new plant in rural Illinois.

Since Japanese automakers manufactured very few of their own parts, most components assembled in their U.S. plants had to be imported from suppliers in Japan. The automakers said they hoped eventually to buy most of the parts in the United States, but aside from generic items, like tires, U.S. parts manufacturers weren't familiar with Japanese designs and had little experience doing business with Japanese manufacturers. In the few instances when Toyota bought from U.S. suppliers, their defect rates and delivery schedules often were worse than those of Japanese suppliers. American parts suppliers were also surprised to discover that Japanese automakers expected them to perform basic design engineering, unlike their Big Three clients, which had always provided detailed blueprints.

As their U.S. assembly operations gained experience, Toyota, Nissan, Honda, and the others encouraged their Japanese *keiretsu* suppliers to open plants in the United States or to form joint ventures with American parts manufacturers. When in 1985 the value of the dollar increased dramatically against the yen, the local manufacture of automotive parts became all the more vital for purposes of containing costs.

The increasing presence of Japanese automakers in the United States yielded an additional benefit. Japanese automakers were able to understand American tastes more profoundly than they had when operations were based mainly in Japan. By the late 1980s, Japanese automotive companies had opened more than a dozen research-and-development laboratories and design studios in the Midwest and in southern California. The technical centers were charged with helping U.S. parts suppliers and the design studios to translate American tastes into models of future vehicles. The goal was to produce a car that reflected precisely what Americans wanted to buy, no longer to be makers of "foreign" cars—something Ford and General Motors had achieved in Europe, where they were regarded as manufacturers of authentically local vehicles.

The roots of Japanese automotive style stretch back to the turn of the century, when Japan adopted the British custom of driving on the left-hand side of the road with the steering wheel positioned on the right-hand side of the car. British imports were popular for the wealthy few who could afford them. Why the British custom was chosen is not altogether clear. One theory holds that since the samurai carried long swords with handles on the right and long blades extending to the left, horse and pedestrian traffic customarily stayed left to avoid the jostling of a samurai's sword.

Until 1926 foreign automakers were selling only a few thousand vehicles a year in Japan, mostly trucks. Afterward, local producers built a few hundred trucks annually. Ford and General Motors sold from 20,000 to 30,000 vehicles through the late 1930s, some fully assembled, some in kits. After the war, with many Japanese suffering from malnutrition and the country's economy in shambles, the need to conserve material and fuel was self-evident. The government, therefore, imposed a tax on vehicles according to engine size, defining a "small" car as any with an engine of less than 1,500 cc of displacement. Automakers responded by designing engines that were as large as possible without qualifying for the next higher tax bracket. Buyers of a car with a 1,499-cc engine, for example, benefited from as much power as possible without paying the higher tax on a 1,500-cc engine.

As private ownership of vehicles increased, the government invoked more regulations to make sure that cars didn't overwhelm the country's limited network of roads. Space for parking was scarce, so car owners had to prove to the government they owned space to park in order to register a vehicle. Municipal building codes required that parking spaces under apartment buildings be no greater than a specified width (which, of course, dictated the maximum width of most cars in Japan). To limit the number of drivers, the government set a licensing fee that in the West would be considered outrageous: the equivalent of several thousand dollars for a driver's license.

Mandatory annual inspections, called *shaken,* became progressively more thorough and expensive as a car aged; after eight or nine years, the expense of *shaken* virtually forced the purchase of a new car.

In order to motorize the country, Japanese industry had to resolve basic issues of mass manufacturing, finance, and labor relations. The deflationary policies of Joseph Dodge, financial adviser to Occupation Headquarters, induced the impoverished country's first recession in 1949. Overnight, orders (mostly for trucks) and revenue dried up. Automakers didn't have the money to meet their payrolls, which forced them to consider firing workers. As Isuzu and Nissan were issuing layoff notices, Toyota, the industry leader, decided to try the strategy of cutting everyone's pay before laying off a single worker. In return for the workers' agreement to a 10 percent wage cut, the company formally promised not to dismiss employees. But the financial squeeze continued, and lenders were adamant that Toyota cut costs. Toyota had no choice. Despite the agreement, Toyota asked for 1,600 voluntary retirements as a last-ditch tactic to prevent bankruptcy. Accepting personal responsibility for these actions, Kiichiro Toyoda, the automaker's president, resigned along with other senior executives. More than 2,100 Toyota workers then volunteered to retire, a quarter of its work force.

The outbreak of the Korean War and further relaxation of economic controls had put the Japanese auto industry back on track. However, a painful lesson in labor management had been learned. From that time forward, automakers tried never to hire more workers than absolutely necessary, so as never again to face the choice of dishonor or bankruptcy. In addition, because the work force was a fixed, rather than variable, cost, Toyota executives now were compelled to regard workers as lifetime assets. Every worker, from research lab to factory floor, had to be a font of ideas, imagination, creativity, and energy. The automaker couldn't afford to hire someone for strength and good work habits alone.

After the labor strife of the early 1950s, the Japanese automakers enjoyed much more harmonious relations with workers than their American counterparts. Until that time, however, the direction and ideology the Japanese union movement would take was unclear. Some factions favored industry-wide trade unions resembling the United Auto Workers. The pattern that finally emerged was the single-company, or single-enterprise, union that included both white-collar and factory workers and dealt primarily with intracompany issues. Without having to face the bargaining power of a nationwide industrial union, managements were able to moderate wages, keeping them far below those paid by their international competition.

Without union opposition, automotive managements could subcontract the manufacture of a great many parts and components to suppliers that paid still lower wages. Temporary workers could also be hired to buffer the upward and downward swings in production, an option likely to be forbidden by an industrywide union. Temporary workers didn't like losing their jobs when production slowed down, but the solution was practical for Japanese automakers, as it allowed them to offer greater job security to full-time workers. The part-timers, meanwhile, constituted a waiting list of applicants for full-time posts when they became available.

Solutions reached by unions and management on issues such as job security were quite constructive by Western standards. Japanese automakers did not promise *never* to lay off workers. But by their actions they demonstrated that only dire circumstances would justify laying off workers; and in that event, executives could not impose such a hardship without first accepting responsibility for the wider failure it implied.

Adherence to this principle was but one of the many reasons Toyota was able to adapt and improve on practices in American factories. When Eiji Toyoda in 1950 visited Ford's Rouge complex on the Detroit River near Dearborn and saw for himself the vast scale on which the U.S. industry operated, he immediately recognized many improvements that Toyota could make on the American model. But he also understood that the improvements that would bring his company into Ford's league would be incremental and require years to reach their ultimate goal. There was no "great leap" that could catch the mighty U.S. auto industry, but millions upon millions of tortoise-like steps, requiring discipline and patience.

Toyota introduced standard work charts for every worker that stated how much time each job should take, how much material should be kept in stock, and precisely what steps must be performed. The "Toyota Creative Ideas and Suggestions System" was initiated with the purpose of improving every activity in the plant. Teams of workers regularly discussed suggestions. An in-house contest to select a slogan in 1953 produced "Good Thinking, Good Products."

Taiichi Ohno, Toyota's manager of final assembly, studied the flow of parts and materials in the factory with an eye toward improved efficiency. Realizing that producing parts faster or slower than they can be processed at the next work station was inefficient, Ohno and others invented numerous devices for smoothing out the flow.

Toyota engineers invented machining processes that stopped automatically if a piece was slightly defective, permitting the operator to perform other tasks while the machine completed its cycle. This process of self-regulating production was called *jidōka*.

From the moment theirs was designated a strategic industry, Japan's automakers believed they must pursue export opportunities. Generating scarce foreign currency was one objective. By exporting, automakers also could maximize utilization of their factories and increase profits. Latin American and Southeast Asian countries were buying Japanese trucks starting in the early 1950s. Following importation of the first two cars to the United States in 1957, Toyota opened a small sales office in Beverly Hills and began appointing distributors and dealers. But exports to the United States didn't reach appreciable volume until the late 1960s, when technical improvements made vehicles like Toyota's Corolla and New Corona as reliable as other small imports.

The worldwide oil shocks of 1973 and 1979 created an unexpected opportunity for Japan. With no domestic oil supply, the run-up in energy prices hit that country's pocketbook hard. But suddenly the Japanese auto industry's smaller, fuel-efficient cars seemed perfect for American drivers, newly sensitized to the realities of saving energy. To say that the oil shocks were Japan's good luck would be overstating the case; but there is little doubt that years of learning how to build small cars (while America stayed hooked on its big models) served Toyota, Nissan, and Honda very well at a moment when they were searching for new customers and markets.

History might have been quite different had Japanese cars been of mediocre quality. For Americans who came of age prior to the 1970s, "Made in Japan" denoted second-rate construction and cheap material. But as Japanese industry improved the quality of its cameras, televisions, radios, and automobiles, American consumers were not slow to perceive the new standards of quality and value in Japanese goods. If these products came as a pleasant surprise to many, it was only because the West had little knowledge or appreciation of Japan's culture or how the distinctive Japanese spirit and sensibility were manifest in the nation's artwork, tools, machines, gardens, clothing, food, and everyday products.

This wasn't always so. At the second international exposition in London in 1862 — less than a decade after Perry's foray into Tokyo Harbor — Japan displayed 600 examples of painting, craft, tools, medicine, and paper. Simple, spare, and elegant, these Japanese goods were regarded, in some respects, as superior to what Europe had to offer at the time. From then on, Japan set out to absorb as much as possible from the West and blend those ideas with its own. The hybrid tradition in Japan began a thousand years earlier, when the country blended its indigenous culture with those of China and Korea. The ancient Shinto religion, for example, was influenced by the teachings of Confucius and, later, by Buddhism from China. Today, many Japanese observe an amalgam of practices and ideas from all three religions.

The Japanese aesthetic has always been shaped by the caprice of nature and the world of spirits, the scarcity of natural resources and a want of space. For instance, appreciation for miniaturization was manifest in the popularity of bonsai, the growing of tiny plants, an art that demands artistic sense and great patience. Everywhere in daily life craftsmanship is used to create wares for stacking, folding, rolling, and other methods of gracefully conserving space. The traditional kimono must be folded correctly so that it will fit within the precisely measured enclosure provided for it in the Japanese home. If a single fold is incorrect, the kimono won't fit on the shelf.

The artistry of everyday life deeply affects the imaginations of architects, draftsmen, and engineers. The country's first transistor radios were a marvel of compactness that gained much admiration in the West. Transistor radios were soon followed by boom boxes, Walkmen, pocket color televisions. It's not hard to imagine a devotee of flower arranging or the tea ceremony employed as a draftsman in a Nissan or Toyota automotive research lab, sketching the design for an elegant version of cupholder, door handle, or folding seat. Many of these items are not Japanese inventions, of course, but clever iterations of Western products in which practicality is imbued with thoughtful, elegant design.

Japanese artisans, craftsmen, technicians—all those who create—are motivated by a strong sense that people wish to possess what's best or finest. In their ethos, things being created should be formed according to the logic of the user, not the logic of the maker. Even on a factory assembly line, a worker gains a strong feeling of self-affirmation, even pleasure, from his activities because *things* are created through his skills. (In the West, according to Judeo-Christian tradition, work has long been understood to be God's punishment for mankind's fall from grace.) The invisible and divine realm of *kami,* the Shinto spirits that inhabit all things and which must be served, are very closely tied to the act of creating objects, no matter how mundane. Japanese workers believe, on some spiritual level, that their tools or machines are inhabited by *kami,* who would be pleased or annoyed by the results of their labors. Every Japanese factory still is furnished with at least one Shinto shrine, where workers can pray or make a small offering to the *kami.*

American criticisms aimed at Japan's economy rarely acknowledge the sharp contrasts between the two cultures, their ways of thinking, believing, and creating. Few Americans learn to speak or read Japanese. More often than not, Americans still tend to reduce Japanese culture to the stereotype of the well-educated, hardworking people who disdain leisure and are determined to become number one at all costs.

Kenichi Yamamoto, who worked his way up from machinist to president of Mazda, made an extraordinary attempt during the mid-1980s to explain his mindset and philosophy of automaking to American auto executives. "My deep desire is to make automobiles that will touch people personally, that will give them pleasure and improve the quality of their lives," he told executives gathered at a University of Michigan retreat in the summer of 1986. To executives accustomed to discussing cars in clichéd terms like "power," "luxury," and "excitement," these were strange ideas. Yamamoto described his goal of creating harmony for the driver, similar to what a person should feel in the living room of a home. Mere function was not enough, he explained. Automakers should address "the tactile, sensual, and psychological relationship between people and environment, even when in a car." Imbued with this spirit, Japanese automakers devoted abundant attention to small but considerate touches for cars: an interior trunk-release latch, a rotating fan for diverting vented air back and forth from driver to front-seat passenger, a narrow strip of pleasant-feeling material inside the door grip.

Japanese automakers, after listening to criticisms of look-alike cars from American customers, endeavored to impart a distinctive personality to some of their cars. Mazda in the late 1980s introduced the Miata roadster as a sentimental tribute to the British sports cars of the 1950s and 1960s. The Miata won a great deal of praise from reviewers as a car that was a joy to drive and was in many ways reminiscent of the Triumphs and MGs of yesteryear. Mazda engineers had obsessed over attaining a precise "feel" for the way Miata rounded a curve, how its engine sounded—they listened to recording after recording of comparative engine sounds—and thousands of other tiny details. (Interestingly, the Miata appreared just a few years after General Motors was forced to shelve its ill-fated Fiero, a two-seater with numerous mechanical difficulties, including a proclivity for engine fires.)

When Japanese luxury sedans like the Lexus LS400, Infiniti Q45, and Acura Legend were introduced, they underscored the growing impression that Japanese automotive engineering and design had surpassed their counterparts in the West. Consumers simply had never seen or driven cars that were so quiet, smooth, and refined. Nor had they ever sat in cabins so elegantly appointed. When pressed to explain what they achieved, engineers explained that their new luxury sedans were the apotheosis of *kaizen,* the Japanese system of constant improvement. Nothing was so good that it could not be better. What looked perfect was not. When American engineers tore down Japanese engines they discovered no technological breakthroughs, only parts that fit together more precisely than they had imagined possible.

Yet to many of the automotive brains who heard Kenichi Yamamoto speak in 1986, such concepts as "harmony" and "sensuality," when applied to automaking, sounded like so much transcendental mumbo-jumbo. American automotive engineers knew well enough how to improve a car's "feel" by the techniques they had always employed, such as improving the grade of materials, "tweaking" details of design and engineering, and ordering upgraded parts from suppliers. Executives from the Big Three didn't need pious lectures from Japan about an industry that they had dominated for so long. But the simple fact was that since 1980, only thirty-five years after Japan's economy had been rendered virtually prostrate, the country was producing more vehicles than the United States.

To older, more chauvinistic Americans and those who knew little about Japan beyond what they saw in war movies, the thought of the U.S. auto industry being overshadowed by Japan's was galling and frightening. These fears were reinforced by the so-called revisionists, like Clyde Prestowitz, Karel van Wolferen, and Chalmers Johnson, who warned that Japan's economic system posed a threat to American living standards. Free-traders, in their view, were naïve. The United States could never prevail against a society as organized and disciplined as Japan's without adopting specific counterstrategies. To view Japan's accomplishments as the result of legitimate, fair competition ignored that country's unchecked drive for world dominance.

But the Japanese regarded their auto industry with immense pride, proof that Japan was finally entitled to join the first rank of the world's industrial powers. A process that had begun a hundred years earlier was now bearing fruit. Japan had set out to catch up to the West, and its cars, more than any single achievement, proved that their goal had been reached. American criticisms of Japan's auto industry and trade policy sounded to the Japanese like the immature cries of a schoolyard bully who finally had lost a round to a smaller opponent and now blubbered that the contest was fixed.

While these thoughts crossed the minds of Japanese, they were not really acceptable to be expressed openly. The Japanese, by nature, are not outspoken. Since the war Japan had regarded itself as a "little brother" of the United States, under its protection and, to some degree, subject to its authority as leader of the free world. In the Shinto culture, to directly contradict a person of higher status, such as a boss or parent, shows lack of respect. However, failing to correct America's misunderstandings also posed dangers for Japan. Americans tended to take silence or equivocation for positive agreement. If American politicians and demagogues felt that Japan was easily intimidated, they might be encouraged to use Japan as the scapegoat for their own economic troubles. Disagreements allowed to fester could lead to intemperate actions.

In 1989 a book appeared in Japan called *The Japan That Can Say No*. Its authors were Akio Morita, the former chairman of Sony, and Shintaro Ishihara, a leading right-wing member of Japan's Liberal Democratic Party. The book's main thesis was that Japan must argue its economic and political case more forcefully and openly to avoid greater conflict in the future. "One of the reasons why U.S.-Japan relations are in such a mess is that Japan has not told the U.S. the things that need to be said," the book argued. The book's separate chapters were based on speeches that Ishihara and Morita had delivered throughout Japan. Though the book was published in Japanese only, bootleg English versions made their way to the United States and were circulated in Washington.

The Japan That Can Say No critiqued America's economic predicaments from a Japanese perspective. Morita, a familiar figure on the American business scene, censured the short-term mentality of American investors. "Japanese think ahead ten years, Americans ten minutes," he wrote. Americans seek quick profits and sell shares as soon as possible to look for better opportunities. Accordingly, corporate managers (except for military contractors) didn't plan patiently for the long term; they didn't invest enough capital in research and development, lest investors dump their stock.

Ishihara, a fiery politician with a flair for rhetorical excess, blamed American racial prejudice and arrogance for the tension between the two countries. Racism, he asserted, was the chief reason why American military planners had targeted Japanese and not German targets for atomic destruction. "Americans behave more like mad dogs than watchdogs," Ishihara said, suggesting that perhaps Japan should withdraw from its security pact with the United States.

As for the imbalance in automotive trade, Morita noted that in 1987 the Big Three automakers themselves had bought 250,000 cars from Japan while selling only 4,000 U.S.-built cars in his country. "They make no effort at all to sell their cars in Japan, and then call Japan unfair because Japan sells too much in the U.S.," he explained. The trade imbalance, observed Morita, stemmed from "commercial transactions based on preferences" of consumers, not government policy. He singled out for criticism Iacocca, who was widely regarded in Japan as America's leading Japan-basher. In an interview for *Sixty Minutes,* Diane Sawyer had asked Morita what he thought about Iacocca's accusations of Japanese unfairness. Morita told her that Iacocca was "a disgrace" and that he was "unfair." "Since this is a program he would be sure to see," he wrote, "I was frank in my statements."

Sony owned extensive operations in the United States, which gave Morita the opportunity to study American corporate behavior at close range. America loved to lecture the world about the rights of individuals, he wrote, but American corporations often treated their own workers shabbily. Executives were lavishly paid and carried themselves in the manner of Hollywood celebrities; at the same time, they treated "workers as mere tools which they can use to assure profits and then dump whenever the market sags."

Sony's operations in California had been hurt by the recession of the early 1970s. But instead of closing their plant and laying off the 250 workers to cut costs, the company had paid the workers to be retrained in other skills, "out of which grew not only a sense of appreciation, but also a real emotional involvement with the company,"

Morita wrote. Sony's employment in California later grew to 1,500. A worker's loyalty to a company was a "formidable asset," engendered by stable employment and the personal example of executives. Instead of whining about trade deficits, Morita said, American executives should learn a lesson from their Japanese counterparts. Japanese executives never dreamed of paying themselves millions of dollars a year. Perks like private jets and "golden parachute" employment contracts that compensated them richly in the event of firing were unheard of in Japan, where business executives flew commercial, carried their own briefcases, and often paid for cars and drivers out of their own pocket.

The Japan That Can Say No, a book written for the Japanese, provoked a stir in the United States. To those who feared Japanese economic power and favored get-tough measures on trade, the book provided more evidence of Japan's latent hostility. Sander Levin, a Democratic representative from suburban Detroit, read a copy of the book into the Congressional Record. He and Ishihara had met amicably in Washington, though Levin insisted to him that Ishihara and Morita had "misunderstood" American racial attitudes. His constituents, he said, were justifiably worried about automotive trade and how it threatened their jobs.

THE HONDA TEAM

LEE IACOCCA'S insistence that Japan was playing dirty didn't change the reality that Japanese automakers were winning over a disproportionate percentage of young buyers, the industry's future. American youth, who once had hot-rodded in Dodge Daytonas and Pontiac GTOs, were aspiring to new dream cars like the Toyota Celica and Honda CRX.

The trend really began with the parents of these younger drivers. Children born during the post–World War II baby boom distinguished themselves as consumers by rejecting their parents' choices of all sorts of products, including cars. Market researchers first detected the trend in the late 1970s, when "Chevy families" and "Ford families" were replaced by families that aspired to BMWs and Honda Accords. The children of these "import-committed" families grew up not knowing much about domestic cars. Incredibly, some had never driven an American-made car. For them, a Chevy was a car you rented when you went on vacation, not a car you wanted to buy. Among older buyers, about 15 percent were buying imports; but in the twenty-five-to-thirty-four age group, the percentage of import buyers jumped to almost 50 percent.

For Detroit, younger buyers were becoming a lost generation. They had to be won back for the sake of the industry's long-term health. Detroit had a dismal record creating so-called import-fighters, which were supposed to attract younger buyers. Ford's Pinto and GM's

Vega were among the more notable flops. GM's Saturn project was an exception. Saturn attempted to demonstrate that a small, fun-to-drive car that appealed to young buyers could be built profitably in the United States. The demographics of Saturn buyers were exactly what GM was seeking. On a relative scale, however, Saturn was too small a venture to help the industry much.

Iacocca didn't need the services of market researchers, nor did he have to attend focus groups in California, to see the lost generation firsthand. Thousands of younger managers and engineers who worked for Chrysler knew all too well that the cars and trucks they were designing, building, selling, and promoting were not the cars and trucks that they — or other young buyers — wanted to own. Naturally, Chrysler workers bought and drove Chrysler vehicles because the company offered them at attractive prices and because parking an import in a Chrysler parking lot was regarded as disloyal and might well hurt one's career. To keep informed about youthful tastes, Chrysler conducted exhaustive research. In addition, the automaker designated a dozen or so younger salaried employees to meet occasionally for the purpose of recommending ideas to senior executives that might help sell vehicles to younger buyers. Chrysler's "youth committee" came up with ideas ranging from what colors were popular among the young to features that young buyers might desire in their vehicles. At the same time, the youth committee had relatively little influence over Chrysler's overall strategy.

When it came to drawing young buyers into showrooms, no automaker during the 1980s surpassed Honda. Time and again, young Chrysler managers pointed to Honda when offering suggestions on how to improve the Dodge, Plymouth, and Chrysler brands. Honda, the organization, was as innovative as Honda, the car. The automaker seemed to listen carefully to its dealers and customers when searching for ways to innovate. Honda's advertisements were hip and grabbed the attention of young readers and viewers. Honda went to extraordinary lengths to make sure customers weren't disappointed with the quality or performance of its vehicles. Honda, moreover, succeeded as an employer: workers at its U.S. factories were quite content relative to their Big Three counterparts, as evidenced by the failure of the United Auto Workers to unionize the Japanese automaker's manufacturing complex in Ohio.

Honda this. Honda that. Iacocca and other top executives—to their befuddlement and chagrin—were hearing the name Honda day in and day out. The same happened whenever they asked the youth committee for recommendations. Hal Sperlich, the only executive who openly professed admiration for Japanese automakers, suggested to Jerry Greenwald in 1987 that it was time for a systematic study of exactly how Honda developed its vehicles, how it organized its company and how that information might be useful in Chrysler's own operations. Fourteen members of Chrysler's youth committee, which included several M.B.A.s, were drafted to carry out a thorough study of the Japanese automaker.

Sperlich's plan was controversial and a bit risky. Until that time, Iacocca tended to paint the Japanese as a nefarious conspiracy between government and business bent on undermining America's economy. Studying Honda as a corporation suggested that its ideas and business practices—not its "Japan-ness" or its government help—lay behind the popularity of its cars.

Over a period of five months the members of Chrysler's "Honda team" met before work and on evenings and weekends. They studied every publicly available book, document, journal, and newspaper story they could find that dealt with Honda. Reiko McKendry, a Japanese-American woman who worked for Chrysler, helped gather and translate material from the Japanese. Honda, it turned out, was quite open about the way it operated and regularly granted interviews; hence the team had an extraordinary amount of material to absorb and digest.

The company founded by Soichiro Honda in 1948 as a maker of motorcycles was at once typically Japanese and quite distinctive. Honda was a visionary and a technical genius. Honda and Takeo Fujisawa, a highly talented businessman, established the organization and business philosophy, which emphasized a youthful attitude and an enterprising spirit. Corporate decisions at Honda were made after long discussions among many people, as was the case at many Japanese companies. Among Japanese automakers, however, Honda best exemplified an aversion to bureaucracy and hierarchy. Soichiro Honda urged his engineers to pursue passionately the research they cared about, to maintain "a youthful spirit," and not to worry too much about making mistakes or about straying from specific company projects.

Honda engineers and product planners spent prodigious amounts of time observing how motorcycles and cars were used by customers. These observations constituted the first step in transforming information into design innovations. Since Honda had been manufacturing cars only since the early 1970s, the company wasn't bogged down by the industry's traditions. Designers of the first Honda Civic spent days in the parking lot of Disneyland, as well as other locations, to gain less obvious insights into the ways cars were actually used by drivers and passengers. Focus groups and market research, used so devotedly by the Big Three, were regarded by Honda executives as unreliable substitutes for firsthand observation and a free imagination.

Sales experts, manufacturing specialists, and designers worked together on Honda's vehicle development teams under the supervision of a leader. The actual number of workers and the time they needed to develop a new vehicle was quite small compared to normal practice in Detroit. After Honda executives helped establish the general concept for a new vehicle, they refrained from meddling in the development team's operation. It fell to the project leader to make sure the team's work remained faithful to the original concept. By contrast with the frequent changes of assignment common to Big Three project managers, a Honda project leader served for the duration, assuming full responsibility. The project leader was endowed with extraordinary authority to arbitrate conflicts that elsewhere would have to be resolved by top executives. As a result, managers of fairly low rank held substantial power and were encouraged to use it; they were freed from the departmental chain of command and routine political jockeying among and between departments.

What Chrysler's "Honda team" learned about product development in 1987 was later published in academic format by Kim B. Clark and Takahiro Fujimoto of the Harvard Business School, who studied

Honda's product development system in comparison to those of other manufacturers. "Companies that consistently develop successful products—products with integrity—are themselves coherent and integrated," they wrote in 1990. The old-style functional organization, with separate hierarchies for design, engineering, manufacturing, and administration were unable to anticipate the desires of future customers. American automakers had recognized this fact and were busily creating committees, liaisons, and other mechanisms to improve communication between the slow, unwieldy hierarchies. Yet American project leaders did not wield authentic power. Their bosses, the executives, still had the final say. Thus, organizations continued to create vehicles that were developed slowly and fell short of customers' expectations.

In a series of meetings in early 1988, members of Chrysler's Honda team presented their findings to their managers, who were impressed with the simple, straightforward rationality of Honda's practices as compared with those at Chrysler. It was hard to deny that Honda was organized to conceptualize, build, and market cars in a stunningly effective manner. Honda's sales had suggested organizational excellence; the study confirmed it.

Honda team members were pleased that their managers were receptive to the findings. It seemed obvious to them that Chrysler could improve dramatically by simply adapting Honda's practices. Such an initiative, however, required the backing of executives and of Iacocca himself. In early 1988 the Honda team presented its findings to Steve Miller and Jerry Greenwald. Both men were impressed, but though they were the two most powerful executives at Chrysler behind Iacocca, only the supreme leader was permitted to tinker with the organizational chart. Chrysler was in many ways Iacocca's personal fief. His judgment, more than anyone else's, had shaped what kind of cars and trucks were rolling off Chrysler's assembly lines, what kind of advertisements were broadcast on television, what marketing strategies the company pursued. Without his agreement, a change in the way corporate power was apportioned was unthinkable.

The Honda team's study proposed a thoroughly new set of practices that would drastically transform the role of senior executives in product development. In the proposed scheme, after these executives defined what type of vehicle to develop, development would be largely mapped out by independent teams that would include all the functional specialists needed to completely engineer a car. At Honda a seven-member *senmukai*—meeting of senior managing directors— defined the "what"; the project teams decided the "how." Changes and suggestions by executives were discouraged. In a company modeled after Honda, there would be no Lee Iacoccas "suggesting" to designers how a grille or a headlamp should look or which engine should be installed in a vehicle.

Honda's management constantly encouraged employees to express themselves and not worry about making mistakes. The members of Chrysler's Honda team were struck by how this philosophy differed from the message sent by their own executives to the workforce. The team had conducted an informal survey of 270 Chrysler workers, asking open-ended questions such as, "What do you think Chrysler's priorities are?" and "What do you think Chrysler's priorities should be?" The most common answer to the first question, the answer given by 41 percent of those surveyed, was "profit." The most common answer to the second question was "quality," given by 30 percent of the respondents, followed closely by "customer satisfaction," given by 21 percent of the respondents.

Indeed, Chrysler's formal objective in 1987 — articulated by Iacocca to top management — was to maximize the value of its stock to $100 per share by 1992. To reach the $100-a-share goal, Chrysler had to boost profit, using short-term tactics like reducing the number of employees, cutting costs by trimming benefits and supplier prices, or boosting the number of vehicles sold to daily-rental fleets. (Another method of boosting share price was to increase the ratio of the price of Chrysler stock to its profit per share. Chrysler stock had been selling at a multiple of about three times earnings, or three times profit per share. Thus, if Chrysler earned, say, five dollars a share annually, the market thought the price of the stock should be about fifteen dollars. But some nonautomotive industrial companies sold for fifteen or twenty times earnings, or more. By diversifying into other businesses, Lee Iacocca had hoped the market would regard Chrysler as a "diversified manufacturer" and assign a heftier multiple for the stock.)

To members of the Honda team, Chrysler's strategy seemed tantamount to declaring that "the way to win the basketball game is to concentrate on the score." Honda, by contrast, seemed to focus on fundamentals: passing, shooting, and excellent physical conditioning. The Honda philosophy emphasized "customer satisfaction" as the company's goal. Members of Chrysler's Honda team discovered that the words weren't just a slogan. Workers from every part of Honda's operations conferred with dealers and owners to find out about problems and fix them, whether that meant redesigning a part or changing a manufacturing process. By Chrysler standards, Honda workers received an extraordinary measure of trust. Honda's philosophy, though difficult to quantify precisely on a chart, translated into a "feeling," a kind of Zen that can't be experienced without "standing on the factory floor and by talking with Honda associates," reported a member of the research team at a presentation to his superiors.

Rather than investing in plants and tools with an eye toward maximizing production during economic booms, Honda adopted a conservative production philosophy. The concept was to forgo maximizing profit during peak economic demand in order to prevent large losses when the economy was soft. Rather than introduce models and nameplates and then withdraw them, Honda had added models and nameplates slowly and deliberately. The penalty for such deliberateness was that Honda hadn't yet entered the booming truck business. On the other hand, Honda had never dropped a nameplate. Chrysler, by comparison, sold vehicles under a hodge-podge of forty-five different nameplates between 1975 and 1987, twenty-four of which had been dropped. Only the New Yorker and the Gran Fury survived for the length of the entire period.

The Honda team noted that the Japanese automaker placed great emphasis on research and development, devoting a substantial amount of its revenue to new technology and product improvements. Chrysler lagged behind significantly, devoting about 3 percent of its revenue to R&D, while Honda invested more than 5 percent. In its report, the team recalled that during the Great Depression Walter Chrysler was particularly proud that "no matter how gloomy the outlook, I never cut one single penny from our research budget. . . . The research of today is what will keep a soundly managed industry alive and healthy five to ten years in the future."

After comparing the two automakers, the Honda team suggested two broad avenues for change. First, Chrysler must stop concentrating on tactics to increase short-term profit and turn instead to improving quality and customer satisfaction. They pointed out that Chrysler's top management incentive-compensation formula depended heavily on short-term profit, rather than quality, market share, or other factors that would spell long-term achievements for the company. "In order to counter this situation," said the group's Reiko McKendry, "we recommend changing the incentive-compensation formula to one which would force us to shift our focus to the long term. Customer satisfaction must become the major component." The bonus pool for the few thousand or so top managers wouldn't entirely disregard profit, but its importance would be reduced. Moreover, *all* employees should receive incentive compensation, she said.

The second recommendation was to switch to cross-functional teams. Department hierarchies should be replaced by development teams that contain members specializing in engineering, manufacturing, finance, marketing, and every other discipline needed to design, build, and sell vehicles. Since Honda's teamwork was directed by top executives who worked in an office without walls, Chrysler should do the same. "This is both physical and symbolic," Ms. McKendry said. "On one hand we believe that a bulldozer should be brought up to the fifth floor of the Keller Building and physically knock down the walls. This action would force our top officers to act as a team more frequently and send a clear and unequivocal message to the rest of the company that things are changing at Chrysler."

Well aware that Lee Iacocca and his compatriots weren't about to give up their perks or their private offices, McKendry acknowledged that "physically instituting a community office room for our top officers may be difficult in the short term." Ultimately, however, "no-walls teamwork must be implemented at the top for the sake of

Chrysler's survival," she said. To reduce rivalry for promotions and in-fighting between departments, Chrysler should study ways to "flatten" the hierarchy. One feature of Honda's personnel philosophy that had impressed the research team was its willingness to promote engineers and other professionals without forcing them to accept supervisory responsibility. A great engineer was able to continue to contribute his specific talent without being forced into the role of a manager for the sake of higher pay and benefits. At Chrysler, status and pay depended almost solely on how many people reported to you.

Honda trained and retrained workers relentlessly, while Chrysler often prevented workers from taking courses relevant to their jobs because they interfered with daily operations. The team recommended course study requirements for every job classification to encourage continuing education. Training should be cross-functional, that is, it should give employees in functions as disparate as product development and marketing the chance to be exposed to all the fundamentals of the automobile business. Finally, Chrysler needed "youthful," vital ideas and, therefore, must recruit young people.

At the end of the presentation, dated February 1, 1988, the Honda team summarized its suggested roadmap for Chrysler's future as simply as possible:

"Developmental Order for Chrysler Corporation"
Create a company that:
1) trusts in its people,
2) believes in the team concept,
3) has a long-term focus on customer satisfaction,
4) and values product excellence and innovation above all other business considerations.

Publicly, Iacocca had plenty of spicy invective for Japan. But inside Chrysler he learned that the Japanese auto industry had answers for Chrysler's troubles. The Honda team's activity was not the only evidence of a deep institutional desire to transform Chrysler's operations. During a trip to South Korea in 1987, Iacocca was astonished by the astuteness and sophistication of the young managers at

Samsung, an industrial conglomerate. "They were asking questions right out of yesterday's *Wall Street Journal*," Iacocca remarked. The Far East had technology, it had financial resources, and now its managers seemed as good as or better than those in America. If Chrysler's people weren't at least as good, the company had no hope of competing. Upon his return, Iacocca asked Tony St. John, head of personnel and labor relations, to develop an executive education program for Chrysler's middle and senior managers. Iacocca could no longer ignore the fact that Chrysler had been staring at its navel while others explored better ways to organize.

St. John called acquaintances at the Harvard Business School, who referred him to the Center for Executive Development, also in Cambridge. Doug Anderson, a former Harvard instructor and one of the founders, agreed to explain to Iacocca, Greenwald, and other senior officers how his organization used education as "a strategic agent for change and competitive advantage." AT&T, Ford, GE, and many other large corporations already had corporate education programs. They consisted of seminars—often held on a neutral site off company grounds and conducted by nonemployee "facilitators"—that permitted managers to discuss business problems frankly, without fear of negative consequences from their bosses. This training attempted to initiate a permanent culture of learning and communication, so that companies might develop rational strategies and tactics instead of groping for direction by trying to figure out what the boss wanted.

Iacocca liked Doug Anderson's presentation. Within a few months Chrysler had committed to a five-year program, at a cost of several million dollars, for training its top managers. Starting in October 1988, the top five hundred managers of the company—thirty to forty at a time—began meeting for six-day seminars at a conference center in Kohler, Wisconsin. It was evident at once to the facilitators that Chrysler was a particularly sclerotic organization, full of taboos and sacred cows, in which people were terrified to speak frankly. As financial conditions worsened, the atmosphere had grown more repressive. But now, at the seminars in Kohler, where stripes and other marks of rank were checked at the door, the discussion took on the air of a dam breaking, and emotion gushed forth. Suddenly managers were able to talk openly about Chrysler's massive internal contradictions, the attempts to bamboozle customers with shoddy quality and with vehicles they didn't want. Sentiments and suggestions were startlingly similar to what had been expressed by the Honda team.

With few exceptions, Chrysler managers felt relief at finally being able to express what had so long been on their minds. Many repeated the same sentiment: Chrysler's officers, including Iacocca, must experience a Kohler seminar, what was becoming known as "truth week." In April 1989 a seminar for the top two dozen officers was scheduled in West Palm Beach, Florida, a site chosen because the center in Kohler was already booked. Iacocca was asked to join the group for the last day only. Miller, Greenwald, Ben Bidwell, Tom Gale, and Dick Dauch attended. The results of the seminar were quite similar to what the senior managers had experienced at Kohler. Officers suddenly felt able to express ideas to one another that they never had before. The effect was stunning. Among other things, they were finally able to admit to one another that they were deeply worried about Chrysler's financial condition. They unanimously agreed that the notion of spending $1.5 billion to build the Chrysler Technology Center, at a time when the corporation was so short of cash, struck them as seriously loony. None of them had had the guts to express reservations about the technology center before, knowing it was a pet project of Iacocca's.

The tech center project provided a perfect example of the dark side of striving always to be a team player. To help officers understand the potential fallout of going along with an idea and keeping one's mouth shut, Doug Anderson recounted the "The Abilene Paradox," a parable that illustrated why information, ideas, and feelings often remained bottled up in corporations like Chrysler. The Abilene paradox opens with Grandpa and his family sweltering on their porch in rural Texas. Grandpa says, "What do you say we go to Abilene and have dinner at the cafeteria?" It's about 106 degrees and no one feels like getting on the bus to Abilene. Nevertheless, one by one, all the family members say, "That sounds like a good idea." Several hours later, hot and tired and full of mediocre food, the family has returned. "I don't know why we just didn't stay home and have cold cuts,"

Grandpa says. Even Grandpa, the one who proposed it, thought the bus to Abilene had been an awful idea. But no one had been prepared to say so. Again and again in hierarchical organizations, managers "board the bus to Abilene" because they are afraid of conflict. They fear that criticizing an idea, especially when initiated by a boss, will show that they're not team players. Employees, therefore, are reluctant to criticize.

When Iacocca showed up for the last day of the seminar, Anderson summarized what the officers had learned. Then he explained the Abilene paradox, substituting the example of the officers' fear that too much was being spent to build the tech center.

"Why hadn't someone said so?" said Iacocca. "We can knock $500 million off the price tag." Iacocca was impressed with the enthusiasm the seminar generated among Bidwell, Gale, and the others. Without a trace of irony, he addressed the group on the last day, declaring in a fiery voice, "Intimidation is out, participation is in. *And I mean it!*"

In the months following the stock-market crash of October 1987, Iacocca had more pressing worries than how to reorganize Chrysler. He had given his blessing to the Honda team. Aware that the literature of business was filled with calls for reform of corporate culture, he was unsurprised by the calls from younger managers for more "openness" and "consensus." These were the fashionable buzzwords of freshly minted M.B.A.s smitten with Japan. He had no intention of bringing egalitarianism to the executive suite, but if his managers thought these ideas would help, so be it. Iacocca also endorsed the executive education seminars that Doug Anderson and the Center for Executive Development would soon be providing for Chrysler middle managers and executives in Kohler. He was, after all, constantly preaching about the importance of education.

Far more pressing to Iacocca, however, was the fact that the U.S. vehicle market had had a good run and the crash now threatened to end it all. He might have lost his insight into the tastes of younger car buyers, but he maintained a keen sense of economic cycles. Bloated by the extra costs incurred by the acquisition of American Motors,

Chrysler would have trouble withstanding the effects of a serious sales slowdown. A falling economy inevitably exposed an automaker's weaknesses.

American Motors engineers and planners, having become Chrysler employees as a result of the acquisition, felt the most vulnerable to any efficiency measures. American Motors employees were the newcomers and had no obvious place in Chrysler's existing structure. Meanwhile, Chrysler engineering managers, proud that their company had survived, wondered vocally what justified keeping American Motors engineers in their jobs — automotive triumphs like the Gremlin and the Alliance? The apparent redundancy and brewing dissension between engineers from the two companies was exactly what Hal Sperlich had feared and was one reason he had voted against the acquisition.

The disagreement over how to handle AMC was one of the factors that set the stage for a showdown between Bob Lutz and Hal Sperlich. Lutz found Sperlich to be a likable sort, if argumentative and sometimes prickly. But at the end of the day, Hal Sperlich was an engineer who would always do Iacocca's bidding. Sperlich had opposed the Maserati TC, the luxury roadster that was supposed to elevate Chrysler's image; he knew its fakery would never work with sophisticated car buyers, but when Iacocca said to do it, he went along. Sperlich controlled design and product development, and he commanded the loyalty of Chrysler's engineering community. His army included many troops from his days at Ford, the foot soldiers who produced the minivan and the variants of the K car. Iacocca had the final say on how a vehicle looked and performed, but it was Sperlich who provided him with his choices.

In the view of Lutz and François Castaing, Sperlich had put too much emphasis on vehicle styling to win customers. To them Sperlich was a metal-bender, a cartoonist in a sense who loved to sketch designs. Sperlich built vehicles from the outside in. For him a car's power and technical performance were not as important as its visual signature. Lutz and Castaing, on the other hand, conceived of vehicles from the inside out. Philosophically, the soul of a vehicle was its power train — its engine and transmission — which in turn established the theme for the chassis, suspension, seating layout, body, and—finally—the styling and appointments the vehicle would contain.

Iacocca's hiring of Bob Lutz in mid-1986 told Hal Sperlich that a contest of automaking philosophies and of corporate power had been set in motion in which there could be only one winner. Whenever the two men were together, the tension was palpable. Iacocca tried to assure Sperlich that he shouldn't feel threatened by Lutz, that there was plenty of turf for both of them. Sperlich didn't believe it; he had seen Henry Ford use the identical technique of creating rivalries to separate the weak from the strong many times, and Iacocca had learned well from Henry Ford. The AMC acquisition had been a major defeat for Sperlich, who had favored spending the money instead on a new vehicle platform. Lutz also had reservations about AMC but voted for it. Lutz was now telling whoever would listen—ever so diplomatically, of course—that Chrysler's cars were losing market share to the Japanese because they were second rate and that Sperlich and the engineers under him were to blame. Lutz also blamed Sperlich for the joint venture with Mitsubishi to build cars in Illinois, a project Lutz thought unnecessary.

That Iacocca had forced Sperlich to build vehicles that were now Sperlich's liabilities in his struggle against Lutz was immaterial. Unless he could somehow find a way to defeat Lutz, his corporate rival, Sperlich was finished. He never found the way.

Lee Iacocca, in a corporate press release dated January 21, 1988, announced that Harold K. Sperlich "has decided to retire after more than 30 years in the automotive industry." Technically, the release was accurate. The truth, however, was that Iacocca had summoned his longtime ally and told him that he was out. Sperlich could stay and collect a salary if he wanted; he could stay on the board of directors; he could run Chrysler's international operations, which were virtually nonexistent. But he wouldn't be in charge of building cars anymore. Iacocca was giving Bob Lutz that job; he and François Castaing were being given the chance to put vehicles in the showrooms that could stop Chrysler's inexorable slide.

The first time Iacocca had fired Sperlich, at Ford a decade earlier, he committed the act under duress. Henry Ford had ordered Iacocca to get rid of his longtime colleague. This time it was different. The deed and the responsibility were Iacocca's alone. Iacocca made sure Sperlich received double the normal retirement benefits he was entitled to. Iacocca played the role of benevolent despot rather than that of a Henry Ford on the rampage.

The result, in any event, was the same. There wasn't room for both Lutz and Sperlich.

The life of the factory shifted from secondary to primary importance in Detroit once every three years when the United Automobile Workers Union and the managements of the Big Three sat down to negotiate wage and benefit scales. To executives, manufacturing was a routine that happened each day and was mostly taken for granted, like water tumbling over a falls. In the early days of automaking, manufacturing processes were the essence of the business, and executives like Walter Chrysler and Alfred Sloan understood them. Since the advent of the moving assembly line and interchangeable parts, there had been no comparable revolutions in the American factory. As early as the 1940s, styling, technical improvement of the vehicles, marketing, and finance had eclipsed manufacturing as fields of innovation. Work on the assembly line was organized much as it always had been. Few, if any, executives of the modern age of automaking had ever worked in a factory or cared much about manufacturing. A talented production supervisor or brilliant plant manager was rarely promoted to a position of influence at headquarters. The accountant who recognized how to remove a penny of cost from a three-cent part was the man to watch. Automotive executives knew that the entire enterprise depended on what came off the assembly line each day — they all had attended lectures preaching the importance of respect for the line workers. Privately, however, the average white-collar manager looked down on the "factory rats" and resented the union's control over their operations. Union rules governed exactly how many minutes each person had to be at work, what kind of jobs they were allowed to do, and even — by virtue of the grievance system — how supervisors could discipline them. To executives, factory workers were slow moving, undereducated, and pampered. They drank too much, many of them took drugs, and the union shielded them from the consequences of their misbehavior.

In an interview in *Fortune* in 1991, Iacocca unwittingly laid bare the Detroit executive's prejudices concerning the denizens of their auto plants. In answer to a question about competition with Japanese automakers operating in rural U.S. settings, he said, "They start with a young work force, no health care, no pensions, which saves them $600 per car. So I should go to Iowa to build a plant and screen the workers to make sure they're young and haven't been on drugs? Do that kind of screening in Detroit, and you won't have anybody working for you."

The executive's stereotype of the autoworker, of course, was hideously unfair. A job on an assembly line at General Motors, Ford, or Chrysler was typically dirty, hot, boring, bad for your back, and often dangerous. The great majority of people working in auto plants did so because the job paid far better than any other work they could hope to find; the hourly wage, plus time and a half for overtime, and fringe benefits like medical coverage with no deductibles and company-paid pensions afforded many people who had little or no education a middle-class living. Most autoworkers grunted, struggled, and sweated nobly under difficult conditions, wrestling with parts that didn't fit properly and machines and tools that were inadequate.

But Iacocca had a point. Gambling and substance abuse *were* rampant; prostitution, loan-sharking, and other sleazy rackets prospered in the aisles and on the loading docks. Chronic absenteeism and arcane work rules, specifying that workers must perform only certain kinds of jobs, meant that automakers were often employing twice as many workers as they needed. An assembler couldn't be asked to clean or drive a forklift. A driver wasn't allowed to substitute on the line. The union always pushed for more workers. Workers paid dues and helped leaders stay in office. Union officials often resisted efforts to make jobs more efficient, ever suspicious that "improvements" on the line were simply disguised attempts by management to squeeze more work from their members.

Ben Hamper, an employee in a General Motors truck assembly

plant in Flint, Michigan during the 1970s and 1980s, gave voice to
the frustrations and absurdities of factory life from the worker's point
of view. Writing in *Mother Jones* magazine and various newspapers,
Hamper introduced himself to the world as "Rivethead," a booze-
swilling, drug-taking product of a broken home who saw little choice
but to follow in the footsteps of his father and grandfather and toil on
the assembly line. An engaging, funny writer, Hamper described life
building Chevy Suburbans—the hours of monotony, the venality and
incompetence of supervisors, and the dedication of automakers to a
single overarching principle: to build as many vehicles as possible.
Hamper wrote about "doubling up," the practice of doing two jobs at
once while a partner read or slept, and then switching with the partner
after an hour or two. At meal breaks many of the workers headed for
the bars, according to his account, returning in an alcoholic stupor
that undoubtedly contributed to accidents and poor quality. Atten-
dance in the plants was chronically spotty, except for Thursdays,
when pay envelopes were distributed. Occasionally violence flared
among workers or between workers and supervisors.

Hamper's writings, which were summarized in a variety of publications, including the *Wall Street Journal* (and later collected in his
1986 book *Rivethead*), embarrassed and infuriated General Motors
and the United Auto Workers union, as well as the vast majority of
autoworkers who showed up on time, put in an honest day's labor, and
faithfully bore the tedium of the assembly line year in and year out.

No one could refute, however, that Hamper's portrait was drawn
from life. His writings described accurately what a significant percentage of workers experienced each day. Some American auto plants
probably were better than the Chevy Suburban line in Flint, and some
worse; but to some extent they all suffered from the same disease, an
adversarial relationship between managers and workers that inhibited manufacturing reforms and permitted behavior that was inimical
to the production of top-quality vehicles. Workers were regarded by
the automakers as little more than automatons, their ideas fit only for
the proverbial suggestion box. Training, instead of being continuous,
was minimal. In any industry that truly dedicated itself to quality and

progressive labor practices, Ben Hamper and many of his supervisors would never have qualified to set foot inside the factory door.

Hamper and Michael Moore, his editorial sponsor at *Mother Jones,* regularly caricatured Roger B. Smith, the chairman of General Motors, as the archetypal auto executive carting home millions in salary and bonuses and caring nothing for workers trapped in the grind of daily production. (Moore produced and directed *Roger and Me* in 1989, a savagely funny film, a mock-documentary film about Moore's unsuccessful attempt to interview the GM chairman about Flint, where residents were coping with shutdowns of GM plants.) As one-sided and unfair as their attack on Roger Smith was, it echoed and reinforced the compelling argument that American business leaders should accept personal responsibility for the failure, as well as the successes, in the factory. If chief executives are prepared to exercise virtual dictatorial power and pay themselves fortunes, then they should be prepared to accept the consequences for corporate failure. When all went well in Detroit, the industry's top executives lived in regal splendor, and their faces graced the covers of magazines. But when plants were closing and auto towns like Flint were ailing, no one was around to to explain.

Chrysler's equivalent of the Suburban line where Ben Hamper once labored was the Jefferson Avenue assembly plant, a cavernous, 3.6-million-square-foot complex on the east side of Detroit. The Jefferson Avenue plant went back to 1907, making it older than Chrysler itself. At times the plant employed more than 5,000 workers, roughly double the population of a typical Japanese assembly plant. Over the years, millions of vehicles rolled off its assembly lines, establishing the plant as a vital underpinning to the city's economy.

Overlooking the river and Windsor, Ontario, to the south, the factory was a rabbit warren of alleys, stairwells, storage bins—places where workers could hide, take a nap, smoke a joint, buy a pistol, put money down on a number, or participate in any of the other numerous rackets that flourished within its walls. Worker health and safety were at risk because of the irregular design of the place and the awkward positions employees were forced to assume to carry out their jobs.

The "pits," where workers stooped to fasten parts under the vehicles, were particularly dreaded because they were often filled with several inches of water. Any worker who was sore or tired or simply in need of some time off could seek one of the "note doctors" who plied their trade near the plant, providing bogus medical excuses for fifteen or twenty dollars each. Little wonder that absenteeism was high and morale low, or that Local 7 of the UAW was constantly battling with plant managers for better conditions, tools, and protective clothing.

For conscientious workers, a typical day's employment at Jefferson Avenue held little sense of fulfillment or pleasure. Lighting in the plant was poor and the roof leaked. Midsummer heat was unbearable. The original wooden floors remained hazardous (though they were much loved by workers because they were easier on the legs than cement). For production supervisors and efficiency experts the building was a nightmare since it was unsuitable for installation of most modern machinery and computers.

Chrysler executives had tried to shut Jefferson Avenue permanently during the 1970s, but opposition from local and state politicians saved it. Sentiment among workers toward their company had always been cautious, at best; supervisors were regarded as management's stooges, interested only in pushing workers to do more, and to do it faster. During the bailout, Jefferson Avenue workers put aside suspicions and supported wage and benefit concessions. Executives hailed workers as heroes in 1981 when the fuel-efficient K cars they were building—the Dodge Aries and Plymouth Reliant—staved off bankruptcy.

Chrysler's oldest plant was an obstacle to narrowing the quality gap with the Japanese auto industry. There was no way that a K car's paint job could ever match that of a Toyota Camry without installation of the latest climate-controlled, computer-driven painting systems. Jefferson workers were still painting with archaic hand-held sprayers. Investing several hundred million dollars to install new painting equipment made little sense in a building whose leaky roof might ruin the cars' finish at any moment or whose wooden floors could barely support the new machines.

Chrysler desperately wanted to close Jefferson Avenue, but moving from Detroit would have been regarded by many as an unthinkable act of betrayal. After all, Detroit's mayor, Coleman Young, and Democratic politicians from around Michigan had sprung to Chrysler's assistance by lobbying for federal loan guarantees. Iacocca did not dare pull the rug out from under these recent allies.

Toward the end of 1986, Chrysler and the city struck a deal: the city donated the land across from the old Jefferson Avenue plant; state and local government extended $200 million worth of tax breaks and other financial help; and Chrysler agreed to erect a new $1 billion state-of-the-art factory for the assembly of an unspecified new vehicle model. It would be the first new Chrysler factory in a quarter-century.

However, without a quantum leap toward better labor relations and work habits, Chrysler officials told the union, the company had no intention of building a new factory in Detroit. The old Jefferson Avenue factory was slated to close sometime after 1987; a payroll that totaled nearly $300 million annually would go elsewhere. With that ultimatum in mind, members of Local 7 voted to approve a so-called modern operating agreement at the new plant that drastically altered work rules, job classifications, and the management of the plant. The number of job classifications dropped from ninety-eight to ten; the union agreed that workers could be organized into teams and that the number of union stewards and company supervisors would be reduced. Instead of 4,500 to 5,000 workers, the new plant would employ no more than 3,500.

As Japanese transplants were locating in rural or semirural locations, where education achievement was high and wages were low, U.S. auto plants generally had been trying to disengage from urban or suburban areas, where aging, less educated unionized work forces commanded big salaries and resisted modern operating agreements. Chrysler had infuriated the town of Kenosha, Wisconsin, and politicians from that region with its decision to close the ancient American Motors plant there. Following the acquisition of AMC, Jerry Greenwald had hinted that Chrysler might operate the ancient Kenosha plant at least through the early 1990s, based on Iacocca's conviction that large rear-wheel-drive Dodges might remain popular forever. But a softening of demand caused Iacocca, Greenwald, and other

Chrysler executives to rethink Kenosha. Chrysler later had to cough up $40 million to repay state grants and to soothe bad feelings over reneged promises.

Smarting from the reversal in Kenosha, the United Auto Workers union learned in 1988 that Chrysler was preparing to try an audacious tactic to evade union wage scales. As other automakers were closing plants or "outsourcing" more and more parts to manufacturers in Mexico, Chrysler devised a plan to auction off its entire Acustar parts-making subsidiary—twenty-nine plants employing about 30,000 workers, or about half of Chrysler's blue-collar workforce. If Chrysler succeeded, a new owner could press for less expensive labor agreements or perhaps attempt to break the union altogether. Then Chrysler would be free to buy parts and components at the best prices available instead of being locked into purchasing from its own subsidiaries.

Predictably, UAW officials gave Iacocca an ultimatum: forget about spinning off Acustar or bear the consequences. Privately, union leaders were furious that the automaker would think of selling workers down the river just a few years after the union had voted to concede wages and benefits in the name of Chrysler's survival. With union leaders muttering bitterly, contract negotiations in early 1988 opened in a tension-filled atmosphere. The bad karma worsened a few days into the bargaining when Chrysler released its proxy statement to shareholders, which listed the salaries of Iacocca and the other chief executives. Iacocca's total compensation for the previous year was $18 million—down from $23 million the year before, but still astronomically high in the view of the union. (Chrysler public relations argued that Iacocca's compensation reflected mostly stock options granted years earlier that were being cashed in, a point that carried little weight with the union.) The automaker's top 2,000 executives received $102 million in bonuses alone, up from $77 million a year earlier. Chrysler's negotiators at the same time were arguing that job security provisions and a profit-sharing plan, already won by General Motors and Ford workers, cost Chrysler too much to give to its own workers.

The most extravagant executive compensation in the U.S. auto industry combined with the treachery of trying to sell Acustar to explode spectacularly in Chrysler's face. Only personal diplomacy by Lee Iacocca managed to mollify UAW president Owen Bieber and to succeed in smoothing the union's ruffled feathers—but not before Bieber extracted a promise that Chrysler executives no longer would be eligible for bonuses unless workers were paid a profit-sharing bonus. In unprecedented fashion the union had won a limit to executive compensation; Chrysler also dropped its plan to sell Acustar.

But management extracted a concession too. Union leaders promised to cooperate companywide, in the spirit of what had taken place at Jefferson Avenue, to urge relaxation of work rules and job classifications at the factories. Chrysler executives understood that unless their plants could operate as Honda's plant in Marysville and other Japanese transplants did, they stood little chance of catching up with the Japanese. In 1988, workers at only six of Chrysler's forty-six factories had agreed to the modern operating agreements. Unless workers at more plants were enthusiastic about the concept, the effect on the corporation would be minimal. With leaders of the international union on board, Chrysler had the chance to persuade more employees that modern operating agreements weren't another ploy to coax more effort from them.

When Lee Iacocca had first thought about segregating Chrysler parts-making operations into a separate entity called Acustar, he had an additional motive beyond that of bundling factories and workers so they might be sold off more easily. He hoped to remove Acustar from the purview of Chrysler's executive vice president of manufacturing, Richard E. Dauch.

As the man in charge of all of Chrysler's factories, Dick Dauch wielded great influence over labor relations, purchasing, product design, and marketing—but especially over the daily routine of plant floors from Saltillo, Mexico, to Newark, Delaware. During the days of the bailout, the startup of the K car, and the introduction of the minivan, Dauch gained a Patton-like reputation for marching through

Chrysler's chaotic manufacturing operations and whipping them into shape so that they consistently met production quotas. Plant managers and their subalterns quaked when Dauch showed up. When Dauch barked, they responded in kind to their subordinates, and so on down the line to the foremen, whose job was the direct supervision of the workers.

Though Dauch was loyal to Iacocca—almost worshipful, in fact—his aggressive, dictatorial methods grated on others. Several executives begged Iacocca either to rein Dauch in or get rid of him. Dauch had little insight into how he was feared and disliked. He made no bones about the fact that he thought he deserved to be a chief executive officer—if not of Chrysler, then of some other corporation. He had hoped to be appointed to the board of directors, but Iacocca, knowing what several other executives thought of his demeanor, kept putting him off. His desire for higher office suddenly became more pressing with the arrival of Bob Lutz, also a man of ambition who made no secret of his opinion that Dick Dauch was the sort of manufacturing executive who belonged in a museum, not in Chrysler's future. Lutz was Dauch's superior in the corporate pecking order due to his appointment to the board of directors. As Lutz gained more corporate power, by virtue of Sperlich's retirement and by his championing of new vehicles and his growing popularity with engineers, friction between the two men increased.

Iacocca liked Dick Dauch. He was fanatically devoted, constantly praising Iacocca in the press. Moreover, Iacocca knew he could rely on Dauch to make vehicles emerge from factories on time, regardless of the obstacles. The toughest plant rats seemed genuinely in awe of him. If a dirty job had to be done, Dauch instantly volunteered. When every other Chrysler executive insisted that the Maserati TC was hopeless and should be written off, Iacocca sent Dauch to Italy to make sure the cars were produced.

But Dauch also created headaches for Iacocca. The more he expanded his turf, the greater the number of executives who complained to Iacocca about his overbearing tactics. A brawny former fullback from Purdue, Dauch controlled his empire much as a tyrannical football coach controls his players and staff, with a combination of tough love and fear. Surrounded by a cadre of assistants—"killer go-fers," as they were called behind their backs—"Coach" Dauch loved to spring surprise inspections on manufacturing departments, quizzing those responsible about quality or productivity while assistants scribbled notes. If a subordinate made a mistake, Dauch excoriated him unmercifully, usually in profane terms and often in large meetings, while the poor man's peers stared uncomfortably at their shoes. His pep talks were highly animated, spiced heavily with vulgarity and often embarrassingly enthusiastic. He was famous for the crushing handshake that he used to establish dominance over whomever he met. Another favorite tactic was to stare for several minutes without blinking when he was talking to someone. He traveled everywhere in cars painted black and gold, Purdue's colors. Strangely, he often described himself as "a people person."

Just as Ben Hamper represented the stereotype of the alienated, substance-abusing autoworker, Dauch personified the hierarchical manufacturing executive who controlled his minions with fear. Indeed, he knew how employees felt about him and cultivated that fear. In certain factories, employees had devised codes using whistles and horns to warn when Dauch was headed their way for a surprise visit.

To make up his own staff of assistants, Dauch preferred to hire athletes who had excelled in football or some other contact sport "where you are willing to risk getting hurt." In all his years, he claimed, he'd never had an athlete quit on him.

A farmboy from Ohio, Dauch had attended Purdue on a football scholarship. (University athletic department records reflect that he actually played very little due to a knee injury.) According to his version of events, GM executives hired him as a trainee after watching Purdue thrash Michigan on the football field. His rah-rah manner served him well at Chevrolet manufacturing during the 1960s; by the age of thirty-one he had become one of GM's youngest plant managers, in charge of spring and bumper works in Livonia, Michigan. In 1976 he was recruited by ex-GM executive Jim McLernon to open an assembly plant for Volkswagen in New Stanton, Pennsylvania. It was a big career break, for he was named a Volkswagen vice president, albeit of a VW manufacturing subsidiary rather than of the parent corporation. By all accounts Dauch threw himself like a demon into the project of starting production of VW Rabbits at New Stanton, accomplishing the task in the astonishingly short period of eighteen months. Before long he was promoted to group vice president for manufacturing.

Lee Iacocca heard about Dauch and asked him to join Chrysler in May 1979, a time when the automaker's prospects were at their gloomiest. Dauch turned Iacocca down. With the knowledge that Iacocca was courting him, he traveled to VW headquarters in Wolfsburg, Germany, where he sought a promotion. Different versions of the story exist, but several insiders agree that Dauch demanded a more important executive position and to be named a director. Volkswagen executives, offended by Dauch's ego, invited him to look elsewhere for employment. Soon after, in early 1980, Dauch joined Chrysler as a vice president and soon was named executive vice president of stamping, assembly, and diversified operations.

In many ways, Dauch's personality and experience were well matched to Chrysler's situation during the bailout. Chrysler had always pushed production at the expense of quality, a prime reason why customers year by year had drifted away. Yet if factories under Dauch's supervision didn't continue to push vehicles quickly into the hands of dealers, the company surely would fail. For the long term — a long term that always failed to materialize — Chrysler would worry about ways to improve manufacturing quality. For the moment it needed cash.

After Chrysler turned the corner in 1983 and money started pouring in, Dauch pressed for better metalworking machinery and the most advanced painting equipment. He understood that computers allowed machines to be far more flexible, and as a result factories might be renovated so as to be able to shift production of new models within days or weeks instead of months, so that one assembly line could accommodate multiple models. New conveyor systems were installed, as were squadrons of robots. Lighting was improved and the floors received coatings of fresh paint. Selling more vehicles and improving the plants raised morale considerably, yet something was still missing: the involvement of the workers.

Workers in Chrysler factories were confused by the conflicting signals they received. Supervisors lectured them constantly about paying attention to quality, because the company's future was in their hands. Modern tools and equipment, they were told, would help them achieve a level of quality that was equal to anyone's. But whenever they spoke up to discuss reasons for manufacturing glitches — a balky robot, substandard parts from a supplier, and so on — foremen told them to shut up and keep working because meeting the day's production quota remained essential to meeting the week's payroll. (As a finished vehicle reached the shipping gate, Chrysler accountants immediately booked the car as revenue to the account of the dealer it was shipped to.) Far from installing the *andon* cords that in Japanese plants allowed any worker to halt the assembly line to correct trouble, in America, stopping the line constituted a cardinal sin.

DREAMS MEET REALITY

RESTRUCTURING CHRYSLER from a Detroit automaker into a New York–based holding company was a strategy designed in part to fulfill Lee Iacocca's dream to return eventually to the East, where he had grown up. His hometown of Allentown, Pennsylvania, where his mother lived, was only an hour from New York City. Keeping a residence on the East Coast was fine with Peggy Johnson, whom Iacocca was courting as plans were being drawn up in 1985 to legally change Chrysler into a holding company.

Reared in North Carolina, Peggy Johnson had been employed as a stewardess for Pan American Airlines and had been living on Manhattan's East Side. Pretty, dark-haired, and single, she met plenty of powerful, wealthy socialites, entertainers, and business executives on the flights she worked during the 1970s, and she was often invited to their parties and social gatherings in Connecticut, Miami, and Paris. This was how she came to be introduced to Lee Iacocca, a powerful Ford executive, at the Palm Bay Club in Miami by then governor of Florida Claude Kirk. After Iacocca moved from Ford to Chrysler, the two kept up a friendship. When he was appointed to head the Statue of Liberty fund-raising campaign, he arranged for her to work on the committee.

Shortly after Iacocca's wife died in 1983, Iacocca and Peggy Johnson began appearing together in public. He was in love with Johnson, if ambivalent about marrying again. He hated coming home after work to dinner alone. Peggy was a lively, witty conversationalist and a whiz at crossword puzzles. And at thirty-five, she was nearly thirty years younger than he; being with her made him feel young. Unlike Lee's late wife, Peggy challenged Iacocca. If he wanted to stay home and she wanted to go out, she argued for her way. Iacocca was accustomed to subservience, not arguments, particularly from his mate. The loss of his wife was very fresh and painful for his two daughters, Kathi and Lia, who were approaching marriage age themselves. The Iacocca girls didn't expect their father to check into a monastery, but the prospect of his marrying a woman closer to their own age — and one who was far more aggressive than their mother had been — provoked in Iacocca's daughters feelings of confusion and resentment.

Peggy Johnson loved Iacocca, and was willing to move to Detroit to be with him. It would take time for Chrysler to build a presence in New York. At his request, she also converted to Roman Catholicism and agreed to forgo having children. Peggy loved big parties, dinners, nights on the town. She enjoyed Iacocca's celebrated friends and was dazzled by late-night get-togethers in Manhattan with the entertainers and big shots who were his pals. Some sixteen months after their engagement in January 1985, they were at last married in St. Patrick's Cathedral, in April, 1986.

After the Statue of Liberty celebration and the weeks together in New York, the reality of day-to-day work returned. Iacocca, by temperament, was actually a homebody; and Detroit was an early-to-bed, early-to-rise factory town. Peggy presumed that as Mrs. Lee Iacocca she would be traveling a great deal and returning to New York often. The holding-company structure, in which Chrysler would be positioned as an automotive/financial-services/high-technology conglomerate with offices in New York, would give them the best of all worlds. They could keep their primary residence in Detroit, as he wanted; but Iacocca's role as chairman of the conglomerate would release him from the tedious details of automotive operations and allow him a good deal of time in New York, something they both enjoyed.

As the couple began discussing where they were going to live, Peggy Johnson began to experience firsthand Lee Iacocca's reluctance to spend his own money. It was something she had noticed a few years earlier, but she had never worried about it. Starting from the time he was a Ford vice president, Iacocca rarely had to reach for his wallet, except on weekends or when doing personal shopping. The same was true for most auto executives who reached the rank of corporate officer. Some relied heavily on the cash that was always available for personal needs, while others were embarrassed or thought it wrong to spend corporate funds for goods and services that most people had to pay for out of their own pockets. Meals, travel, personal services like massages, haircuts, laundry, books, magazines, theater tickets, flowers, gifts, shoeshines, even clothes were often covered for an officer of an automaker. Before every business trip, high-ranking auto executives customarily received hefty envelopes of cash for incidental expenses.

After thirty years of living on Ford and Chrysler expense accounts, Iacocca had little experience deciding when and how to spend his own money. If he was in a hotel and wanted a magazine or newspaper or something else for himself he typically would pick it up, walk off, and expect one of his assistants to pay for it. Once, when Lee and Peggy traveled to Europe on a vacation, Iacocca discovered on arrival that he had forgotten to bring any money. He literally hadn't a cent in his pocket. Years of relying on assistants to take care of financial arrangements had rendered him incapable of handling such things himself.

Behind Iacocca's back, his subordinates joked about their boss's cheapness: "If you have lunch with someone who looks like Iacocca and sounds like Iacocca," they would say, "rest assured, if he offers to pick up the check, it's not Iacocca." Some of those who knew him were amazed or miffed that Iacocca regularly ignored their requests for donations to the charities and schools on whose boards of trustees they sat, even though he was raking in tens of millions of dollars, more than any auto executive. (Before his autobiography was published he pledged the royalties to the research effort against diabetes, as a memorial to his wife, probably never dreaming they would amount to a small fortune.)

He was anything but shy when it came to spending corporate money on entertainment. Every Christmas, for example, Iacocca hosted a sumptuous black-tie feast for corporate officers and spouses at the Bloomfield Hills Country Club. Most Chrysler officers coveted an invitation to one of these rare occasions when the company's maximum leader revealed himself as an ordinary person and mingled with them in a relaxed setting. Others dreaded the evening, which struck them as stiff and formal. Entertainment often included crooning by Vic Damone or some other Iacocca pal. At the end of the evening a beaming Iacocca would be presented with an expensive gift, usually worth several thousand dollars. (One year it was jackets from his favorite tailor, another year a golf cart.)

Several of the officers were embarrassed by the grandiose scale of the event and the impropriety of Iacocca hosting a lavish tribute for himself, particularly during the years when Chrysler wasn't doing well, due in large measure to some of his management miscues. The party, which cost tens of thousands of dollars, came out of Chrysler's coffers. The gift to Iacocca was a particular source of discomfort: a week or two after the party, every officer received a bill amounting to several hundred dollars to cover Iacocca's gift. None of them was asked whether he wanted to give Iacocca a gift; it was merely presumed that they wouldn't mind being assessed. The party acquired the nickname "the Coleman Young fundraiser," a reference to the mayor of Detroit's notorious fund-raising dinners for which city officials were obligated to buy tickets.

Living in Iacocca's home in Bloomfield Hills brought discord in the Chrysler chairman's relationship with Peggy Johnson. In deference to their late mother, Iacocca's daughters didn't want their father sharing the master bedroom with his new wife, so Peggy agreed to move into what had been a maid's room downstairs. She, of course, was insulted and angry and let him know it. Finally, Iacocca acceded to Peggy's wishes and they bought a new house, a Georgian mansion once owned by Henry Ford II in Grosse Pointe.

The thought of moving to Grosse Pointe—not to mention living in a mansion formerly owned by the Fords—made him nervous. He was Bloomfield Hills, home of the automotive industry's new money. Grosse Pointe was a section of town inhabited by the Fords and Dodges and Fishers, the industry's old money. The Fords and much of Grosse Pointe had never forgiven Iacocca for the public vilification of Henry Ford II in his autobiography; and he had never forgiven the Fords for being white Anglo-Saxon Protestants who didn't think him worthy of running the Ford Motor Company. He worried that Grosse Pointe society would close ranks behind the Fords and blackball him from the country club. Before moving into the new Grosse Pointe home, Peggy told Iacocca she wanted to redecorate. The thought of spending tens of thousands—perhaps hundreds of thousands—of dollars of his own money on new furnishings drove him wild. He roared at her: Why can't you just shop at a furniture store? She couldn't fathom why a man who was worth somewhere between $50 million and $100 million was unwilling to decorate a house properly. Editorial cartoons started appearing in the *Detroit Free Press* and *Detroit News* lampooning the newlyweds' anticipated move to Grosse Pointe. When a cartoon appeared with the caption: "There goes the neighborhood," he had had enough. Iacocca ordered Peggy to get rid of the house. The real estate agent was able to find a buyer in one day.

Peggy Iacocca, of course, saw a side of her husband that the public didn't. She recognized that Iacocca desperately needed the trappings of power and symbols of status to assuage the lurking insecurities he felt on account of his family's immigrant background. As she saw it, his anxieties about WASP society rejecting him might diminish greatly if he just confronted them. She, a woman with no social standing, had opened herself to people in Grosse Pointe; and they had responded very kindly—at least it seemed so to her—assuring her that the Iacoccas would be accepted. If he would just relax a bit, she was sure they would respond the same way to him.

Peggy encouraged the spirit of reconciliation urging her husband to bury the hatchet with Henry Ford II. His obsession with Ford was an emotionally debilitating wound. His resentment of Ford had also colored his business judgment, leading him to underestimate the public's acceptance of the Ford Taurus and the massive changes that its success portended for automotive styling. In September 1987, Ford was hospitalized in Detroit, dying of heart disease.

You don't realize what a favor Henry Ford did for you, Peggy Iacocca told her husband. *Getting fired from Ford brought you to greatness. You're richer, more famous and more influential because of Henry Ford. Thank him. Go to him, tell him there are no hard feelings.*

Not surprisingly, hearing this advice from his wife about a man he hated only infuriated Iacocca. Perhaps the sixteen-month engagement, particularly long for a man in his sixties, was a sign that this marriage wasn't meant to be. Eight months after their wedding, Iacocca's attorneys served her with divorce papers. The marriage was dissolved in November 1987.

As is so often the case in feudal warfare within automobile companies, the defeat of an important chieftain triggered the exile of his supporters, the discrediting of his policies, and the supremacy of a new leader and assistants with a fresh set of ideas. The forced retirement of Hal Sperlich in early 1988 was followed shortly thereafter by the retirement of Jack Withrow, vice president for engineering, and some other senior executives who had risen with Sperlich to run Chrysler's product development and engineering activities. Sperlich quickly found a job running Pulte Homes, a large, publicly owned homebuilding concern. Withrow and most of Sperlich's assistants were offered employment by automotive suppliers, companies only too happy to make room for executives who might help them sell components to the Big Three.

With Bob Lutz now in charge of the broad area that included automotive design, development, and manufacturing activities, Chrysler engineers figured, correctly, that Lutz's supporters would soon be moving into key managerial posts. François Castaing, the Frenchman who had come over from American Motors in the acquisition, was a logical candidate to replace Withrow as vice president in charge of engineering. Castaing, as vice president of Jeep and truck engineering, had reported to Lutz, who had been in charge of Chrysler's truck business. The two men had first become acquainted when

Chrysler executives visited American Motors technical laboratories during the first weeks after the acquisition was announced. Lutz showed great interest in Castaing's explanation of AMC's vehicle projects and future designs. After the merger, Lutz invited Castaing to a long dinner at the elegant Whitney restaurant near Wayne State University. The two men chatted in French about their philosophies of automaking, comparing ideas and experiences, gaining a feel for whether they could be allies.

Castaing and Lutz hit it off. They held similar views about vehicles, namely that great cars and trucks had to be created around central, unifying themes that were based on how they were powered. A vehicle's soul was defined principally by its engine and transmission—the power train, or power pack. Likewise, for a vehicle to be great, the body, chassis, mechanical systems, electrical systems, and appointments all had to suit the power train as well as blend well with one another. Chrysler's vehicles, by contrast, struck them both as a hodgepodge of unmatched systems and components cobbled together on the basis of what was cheap or available, the half-baked offspring of product planners who possessed little technical sophistication, and of executives who believed that marketing savvy could make up for a lack of mechanical integrity. Chrysler's midsize Dakota pickup truck, for example, sprang from the marketing determination that customers might pay for a truck that was bigger than compact but smaller than full-size. The "midsize truck" concept determined the vehicle's exterior dimensions. Engineers and stylists then were told to make do with available parts and components. The Dakota, therefore, was powered with the only Chrysler engine available, a V6—despite the desire of most truck owners for the power of a V8.

Marketing and engineering did not have to be mutually exclusive, Lutz and Castaing agreed. The disciplines were best served when they afforded one another their proper due. Accomplished engineers (like Castaing) and talented marketers (like Lutz) could create vehicle models that attracted and pleased customers as long as senior executives (like Iacocca) and finance men (like Greenwald) didn't constantly interfere with projects and constantly change their minds. The philosophies of Lutz and Castaing were closer to the European way of designing cars, as exemplified by BMW and Mercedes, and not unlike those of Japanese automakers like Honda. Chrysler's system for creating vehicles was similar to the mass-marketing methods of General Motors and Ford, which wasn't surprising since Iacocca, Sperlich, and most top Chrysler executives had spent most of their careers at Ford.

Lutz's promotion was followed soon after by the elevation of Castaing to vice president of engineering. Castaing's advancement sent a warning to loyalists of Sperlich and Withrow, who were sprinkled throughout Chrysler's engineering community, that substantive changes in policy were coming. A new broom had been appointed and was likely to sweep through the engineering departments.

Lutz understood that he would need a large number of supporters to overcome the institutional inertia he was likely to face from engineering. Tom Gale, a highly respected industrial designer who had made career transitions through engineering to pure design, was pleased with Lutz's ascent. Gale knew that he, Lutz, and Castaing were on similar wavelengths and were all eager to build cars and trucks that represented a departure from what Chrysler had been building. Gale supervised Chrysler's designers, who as a group felt stifled by Iacocca's demands for chrome trim, fake wire wheels, and vinyl-padded roofs. Gale and designers like John Herlitz and Trevor Creed—those responsible for creating a vehicle's "look" on both the exterior and interior—had yearned to explore aesthetic themes, lines, fabrics, and materials that appealed to younger and more sophisticated buyers. The promotions of Lutz and Castaing signaled that they might finally get the chance.

Like Lutz and Castaing, Gale was a certifiable car nut. He had grown up in Flint, where his father had worked for Buick. After graduating from Michigan State in industrial design he joined Chrysler fresh out of school. Just as Castaing's automotive sensibilities were essentially European, Gale's were purely American. Lutz and Castaing admired Ferrari, BMW, and English roadsters. Gale loved '32 Ford hot rods and drag racing. He belonged to a breed that loved to cruise up and down Woodward Avenue in Detroit in old Chevys and Mustangs and Dodge Daytonas.

At Lutz's suggestion, Castaing spent more time with Gale, getting acquainted and learning Chrysler's unwritten operating and personnel manual. Gale invited Castaing for a tour of the design "dome" in Highland Park, to inspect prototypes of the Dodge Spirit and Plymouth Acclaim, the so-called A bodies on K car platforms that were to be introduced in early 1989. Castaing was also shown models of the Chrysler New Yorker and Chrysler Imperial, the so-called C bodies on K platforms, models that were to be introduced a year hence.

Though Gale was too diplomatic to say so directly, Castaing got a strong sense from Gale's apologetic air that he was far less proud of these creations than a design chief ought to be. It was plain to see that the cars bore Iacocca's heavy stamp. The Spirit and Acclaim, Chrysler's midsize answer to Accord and Camry, were boxy and undistinguished, even plain—toasters on wheels, as the designers called them. The New Yorker and Imperial featured the slab-sided styling, chrome molding, and vinyl roofs that sixty-year-olds would buy eagerly if they were priced low enough. Castaing realized that Gale's hands had been tied. His designers were designing according to the dictates of Iacocca and the product planners and marketing executives who were second-guessing what Iacocca wanted.

Lutz's mandate to realign the development of new cars represented an increase in his corporate influence and prestige. But Iacocca, ever cautious not to promote a subordinate he could not control, was careful not to arrogate to him too much power. Iacocca transferred Lutz's responsibility for trucks to Ben Bidwell and Lutz's responsibility for international operations to Jerry Greenwald. Instead of bestowing on Lutz the title of president of Chrysler Motors, which had been Sperlich's, Iacocca named him "co-president" along with Bidwell, both men reporting to Greenwald.

This awkward managerial structure, meant to keep executives directly below him slightly off balance, also fulfilled Iacocca's love of "blood sport" in the office. Despite his frequent lectures about the need for executives to work as a team, Iacocca now and then loved throwing a juicy steak into the ring—in this case, the title of president—unchaining the dogs, and letting the strongest of them assert his dominance. If Lutz wanted the promotion, he would have to fight

Bidwell for it. Iacocca had hated it when Henry Ford employed similar tactics to limit his power; but since those days, he had either forgotten how much of his own energy had been wasted fighting Ford, or he may have begun to identify with whatever prompted those managerial practices. In any event, Bidwell and Lutz, never great friends or allies, immediately recognized that Iacocca was attempting to sic them on each other and tried to avoid unnecessary conflict.

Chrysler, they knew, couldn't afford the luxury of intramural squabbling right now. Except for the minivan, nothing Chrysler built was selling profitably. The Japanese were gaining market share by the day, and the Reagan administration saw no justification to make conditions easier for Chrysler by tightening customs and other import regulations as Chrysler wanted. Only better vehicles could save the day.

Lutz, Castaing, and Gale—solidly in agreement that better vehicles had to emerge pronto—reviewed the projects that Sperlich had started. They were simultaneously heartened and worried by what they saw. A new sport-utility vehicle that had been developed by American Motors, code-named Z/J, was in pretty good shape from an engineering perspective. Styling wasn't needed for a while because the Jeep Cherokee sport-utility vehicle was selling well. A spruced-up version of the minivan, with a slightly modified exterior and interior but similar mechanical specifications to the current model, was also in the works. Chrysler had dominated the minivan market since bringing out the first true minivan in 1984. Ford, General Motors, Toyota, and other automakers didn't attract minivan customers in the same number as Chrysler; but they were steadily zeroing in on features that had made the Dodge Caravan and Plymouth Voyager successful, such as low step-up height for women drivers and carlike feel and handling, and were bound to come up with formidable alternatives eventually. Chrysler engineers were attempting to head off competition with refined styling, airbags, and features like innovative built-in child seats, that would obviate the ungainly removable child seats that had become a nuisance of yuppie parenthood.

A great deal of engineering had been completed on the new family sedan, code-named L/H, the much-needed replacement for the K car.

The L/H design, however, looked all wrong to Castaing. Sperlich and Withrow had decided to mount the engine transversely (east–west), with the drive mechanism directly over the front wheels, rather than longitudinally (north–south), thus precluding the possibility of rear-wheel-drive luxury or sport versions, something Lutz and Castaing thought essential. Under the direction of Sperlich and Withrow, the L/H had also grown heavier than its target weight, making it too slow. Product planners had added all sorts of features and options they believed the public demanded, like four-wheel-drive. Every option added weight. At first, the solution was thought to be a bigger engine. A bigger engine, however, would require a new transmission; and a bigger transmission would force a change in architecture for the whole vehicle. The process was turning into an expensive, time-consuming spiral: engineering problem, followed by compromise, followed by new engineering problem, and new compromise—all of which had delayed the L/H and heightened Iacocca's desire to replace Sperlich with Lutz.

Wall Street security analysts, with their eye on Chrysler's sinking share of the car market, began wondering aloud in 1988 whether Chrysler could survive the next cyclical downturn in the U.S. economy. Chrysler attempted to reassure the investment community, pointing out that the automaker planned to spend roughly $15 billion to modernize or to replace most of its cars and trucks during the next five years.

The plan was ambitious and would strain Chrysler's human resources. Due to the sheer amount of engineering required—and the industry's belief that the introduction of new vehicles should be staggered to gain the maximum public exposure on introduction—Chrysler's future projects had to be put in sequence. In any case, Chrysler didn't employ enough engineers to work simultaneously on the minivan face-lift, the Z/J sport-utility, and the L/H family sedan. Because the minivan was the single most important model and plans for its face-lift had progressed to an advanced stage, Iacocca unilaterally decided that the minivan must be finished first. Next would be the Z/J in early 1992—one sport-utility model for Jeep dealers and a second for Dodge dealers. Then came the L/H in mid-1992. Afterward

Chrysler needed a new pickup truck, followed by a new subcompact car, followed by a new compact.

In many ways the most important project was the L/H, to be completed by July of 1992. The new family sedan was to be sold in one version for Jeep and Eagle dealers, one for Dodge dealers, and in three Chrysler versions, including two stretched luxury sedans—a bid to regain ground conquered by Ford Taurus, Honda Accord, and General Motors midsize models.

Lutz, Castaing, and Gale first had to convince Iacocca to scrap the previous L/H concept, the one developed by Hal Sperlich and his cohorts. Gale suggested that the L/H borrow design themes from a concept car called the Lamborghini Portofino, built by Chrysler designers. A concept car is a one-of-a-kind running model intended to force designers to stretch their imaginations and to permit potential customers to express their tastes. The layout of the rakish Portofino, with abbreviated front end, engine in the rear, and disproportionately large cabin, embodied a basic design theme that Gale wished to transpose to the L/H: a smaller, lower front end and bigger cabin. The direction of Gale's thinking indicated a radical departure from the look of Chrysler's other vehicles.

Castaing suggested that Lutz and Gale base the engineering of the new model on Renault's Premier, a midsize sedan that AMC had introduced just as it was being bought by Chrysler. The Premier had also been sold as the Dodge Monaco but had failed to attract customers. The exterior styling of the Premier was too squared off for the new L/H but the architecture did lend itself to a redesigned midsize sedan. From his background with Formula One racers, Castaing was familiar with the engineering issues inherent in moving the driver forward over the front wheels in a mid-engined car. As the driver moved forward in conventional automobiles, the engine compartment had to shrink, forcing all sorts of engineering compromises: the radiator had to be smaller, as did the front suspension struts. Engineers from entirely different disciplines, from electronics to engines, had to cooperate to make sure the design of one part didn't compromise the overall concept.

Assuming Iacocca would accept the new L/H design, a second

dilemma remained which demanded immediate solution: How on earth could the L/H be completed in three-and-a-half years, in time for Iacocca's July 1992 deadline? The Big Three almost always needed at least five years to engineer a new model "from the ground up," in Detroit's lingo. It's true that Ford had managed to bring the Taurus out in four, but GM bumbled for nearly seven years before introducing all the versions of its GM-10 family sedans, the Chevy Lumina, Buick Regal, Olds Cutlass Supreme, and Pontiac Grand Prix. As a result, GM's cars looked dated by the time they were introduced.

Castaing was nervous about the time constraint. Still, having served as chief engineer on Renault's Formula One racing team, he recalled how a small number of highly motivated individuals working as a team had achieved incredible feats of technical prowess and logistics on rigid deadlines. When the starter's flag came down, every person and every piece of machinery had to be ready. Team members learned to pull together, to compromise, to put petty jealousies aside. Following his recruitment from Renault, Castaing had introduced the idea of organizing engineers as teams at chronically cash-starved American Motors. Hence, American Motors had managed to complete multiple projects with fewer than 700 full-time engineers. Engineers acquired multiple skills; they were organized in teams to maximize efficiency and they cooperated with one another. There were few bureaucrats at American Motors because there was no money for their salaries.

Lutz and Iacocca had been impressed by how the tiny AMC technical community had brought out as many vehicles as it did. It was a smaller company, to be sure, but 700 was an unbelievably tiny number of engineers compared with Chrysler's 6,000. When the two automakers merged in 1987, Iacocca had permitted Jeep engineering under Castaing to absorb Dodge truck engineering as an exercise to see whether Jeep's spartan habits might rub off on Dodge.

Chrysler's payroll, pared to the bone during the bailout, had since grown pudgy with all sorts of departments and staffs. Engineers with specialities as narrow as bumpers or door handles were assigned as needed to vehicle projects from departments to which they belonged. Since specific engineers were often needed on more than one project, managers negotiated the assignments, inevitably touching off bickering over who did what, when, and for how long.

Certain that the departmental structure, with its entrenched bureaucracy and endless politicking, would prevent the L/H from being completed on time, Castaing suggested to Lutz that Chrysler's engineering be reorganized. Instead of the department structure, most engineers would be assigned to project teams, called "platform teams," consisting of 700 to 800 members each. Serving together on these self-contained teams would be engineers, designers, managers, marketers, and manufacturing specialists — very much the same type of organization that Honda favored. The minivan would have a team, as would the Z/J, the L/H, and future projects like the new T-300 pickup truck. Later on, teams would be formed for the P/L subcompact and the J/A compact cars.

A thorough reorganization of engineering was a plan fraught with peril. General Motors attempted to streamline its engineering groups in 1984 under the guidelines of a study by McKinsey and Company. The McKinsey study was an attempt to break down the fiefdoms within GM that had slowed vehicle development. But McKinsey vastly underestimated the institutional resistance to change. In the resulting confusion, many GM engineers sat around for months with no clue as to what they were supposed to be doing. The long delays seriously compromised the GM-10 and other new GM models. Lutz and Castaing had to be sure that Chrysler's new organization was easily understood and readily acceptable. A delay like the one GM had experienced was potentially fatal.

The first matter at hand was to secure Iacocca's approval of the new L/H. Tom Gale's designers ordered a crude mock-up of the L/H approximately in the proportions suggested by the switch of the engine from a lateral to a longitudinal position. The front of the car was compact and the windshield was raked steeply. The wheels were pushed out toward the corners, giving the L/H a rather aggressive stance, described as "cab forward," because the passenger cabin was pulled farther forward over the front wheels. (Japanese automakers were experimenting with a similar type of layout at this time.) Iacocca often preached that to be two years ahead of your customers in car design—or two years behind them—was to invite bankruptcy. "Cab forward" was a strange and powerful look, low and a bit squat, something that Iacocca surely would not be accustomed to and might well reject.

Chrysler's product-planning committee consisted of the top twenty or so executives, representing finance, marketing, engineering, manufacturing, and other relevant areas. In January 1989 the full committee convened in the design dome to decide whether to pursue a new design for the L/H. The only committee member who really mattered, of course, was Iacocca. Agreeing to backtrack on the L/H meant a few years' effort and hundreds of millions of dollars down the drain. But to press forward meant bringing out yet another vehicle with little chance of selling strongly or reversing Chrysler's dreadful image. The presentation made by Lutz, Gale, and Castaing was persuasive. The time had come to break from the past. If Iacocca understood how the new L/H repudiated everything he liked in a car, he didn't let on. The chairman cast his vote and nodded his assent.

Less than a month later, Lutz and Castaing invited the automaker's 6,000 engineers to two huge meetings, held on successive days, to explain how they were to be reorganized. At a packed auditorium in downtown Detroit, Castaing explained that most engineers were to be assigned to teams organized around a single car or truck platform. They were about to get the chance to design better vehicles than they had ever thought possible, and in less time than any U.S. car company had ever done so. Instead of reporting to a departmental boss, most engineers would report to the leader of their platform team. They would be encouraged to broaden their skills, to share information, to find creative solutions that didn't require meetings, memos, or any other time-killing, initiative-killing bureaucratic devices.

To those who understood and favored the changes that were sweeping business organizations everywhere, what Castaing described in the meeting was a vindication. But to some loyalists of Hal Sperlich, the reorganization symbolized the imposition of the new regime of Lutz and Castaing, something they had not wanted and had no intention of accepting without a fight.

The cool reception given to the Dodge Spirit and Plymouth Acclaim at their introduction in January of 1989 confirmed the instincts of Lutz, Castaing, and Gale. Floundering engineering and design had caused Chrysler cars to drift badly off the mark. Gale and his designers had dreaded such a result when they completed their final drawings and models of the Spirit and Acclaim nearly five years earlier. The compact Spirit and Acclaim were meant to capture young professionals who favored Japanese offerings, though their homely looks left little chance of that: the new models were woefully behind the times in appearance, branding them as cars for fogeys or nerds. Publicly, Lutz could always find something nice to say about the cars. He had to sell them, after all. Privately, he was appalled.

Reviewers muttered the truth: the Spirit and Acclaim were warmed-over K cars. Perhaps the most cutting judgment was delivered by David E. Davis, one of the industry's most authoritative reviewers and the editor of *Automobile* magazine. His opinions were taken very seriously by auto executives, most of whom knew him on a first-name basis and regularly solicited his thinking. Executives knew that a significant number of buyers bought cars based on what Davis and a few others of his stature recommended. "Chrysler is still doing it with mirrors," declared Davis, "making different cars with the same pieces." Chrysler engineers and marketers, willing to concede how retarded the K car models had become, sardonically referred among themselves to those drivers who bought K cars as "PODS," an acronym that stood for "poor old dumb shits."

The year 1989, it seemed, was to be a year of automotive disasters for Chrysler. No sooner were the Spirit and Acclaim panned than the Chrysler TC by Maserati was introduced to automotive critics. The scorn was unanimous. The luxury roadster, which, not surprisingly, resembled a Chrysler LeBaron—it shared many of the LeBaron's components—had taken five years to complete, was priced at $30,000 and literally shed parts as it rolled down the highway. Iacocca and his old pal from Ford days, Alejandro de Tomaso, had agreed back in 1984 to create the TC as an "image builder," but mismanagement and squabbling among Chrysler and Maserati engineers resulted in delay after delay.

Lutz at one press conference mockingly referred to the Chrysler TC as "a hardy perennial," since the company kept promising that the car was about to be launched only to scrap the introduction until the following season because of "technical difficulties." The premise for creating the TC was more troublesome than any of its technical problems. Iacocca still believed that components from parts bins, welded together and given a prestigious name, could rank as a bona fide entry in the highly demanding European sport-luxury market. (In 1983 he had considered manufacturing a knock-off of a Mercedes roadster based on a Plymouth Reliant—until he was talked out of it.) That sort of marketing chicanery might have worked in the 1960s, but in the 1980s there existed a great many technically sophisticated automobiles competing on their merits. Pretty sheet metal with a fancy name didn't make it. Buyers who spent $30,000 for a car expected specific features and could spot a fake when they saw it. The TC wasn't sporty, it wasn't luxurious, and it was only nominally European. A genuine European luxury car could be purchased for not much more than the Chrysler TC, reviewers pointed out.

Iacocca refused to accept responsibility for the TC, a program that made no sense to anyone else in the company. In his view, it might have worked if his marketers had "positioned" it properly. That failure he blamed on Joe Campana, vice president of the Chrysler-Plymouth car division. Chrysler dealers sold a few thousand TCs and then wrote off the humiliating foray in the hope that nobody would remember the car or the expenditure of several hundred million dollars.

The third and most humiliating debacle of the year was the introduction of the new Chrysler Imperial. Iacocca badly wanted an entry into the traditional luxury market, dominated by Cadillac and Lincoln. The economics of the traditional luxury market were quite attractive, since the variable costs of building a luxury car—bigger engine, premium seating, leather, high-grade stereos, and so on— were only *somewhat* greater than a standard model, while the price customers were willing to pay was *much* greater. In other words, if the gross profit from selling two $17,000 cars was, say, $5,000, the gross profit from one $35,000 car might be twice as much. The financial trick was saving engineering costs by using the same chassis from the $17,000 car on the luxury model. Thus, General Motors used many of the same mechanical components for the Cadillac Deville and the Buick Park Avenue; Ford used the same underbody for the Lincoln Continental and the Ford Taurus.

A primary attribute of a traditional luxury car like the Deville or Continental was size. Luxury buyers were usually older and wanted plenty of room for four or five adults to sit comfortably. Chrysler's K chassis was too small for a luxury body, so Chrysler engineers stretched it longer. (Widening the K platform was out of the question, since that would have required all sorts of new mechanical components, adding greatly to the cost.) The Imperial's length afforded rear passengers room for their legs, although shoulder room was tight. Unfortunately, lengthening the car without making it wider also made the car less stable on turns.

Imparting a luxury image to the Imperial was accomplished with a premium vinyl roof, leather upholstery, slightly better paint, wire wheels, a better sound system, electric seats, and fancy chrome decorations, or "jewelry," for the hood, the trunk latch, and other spots on the body. The result was a narrow K sedan, lacking power from its smallish V-6 engine, that looked like a fancy version of the Reliant from the early 1970s. Iacocca believed that by pricing the Imperial a few thousand dollars below the Deville, he might attract bargain-hunting luxury buyers. Potential buyers were those old enough to remember the big Imperial of yesteryear but not savvy enough to read the enthusiast magazines.

The plan didn't work. The Imperial was introduced at the Algonquin Hotel in New York in November 1989. Ominously, the executive chosen to present the car was Joe Campana, the man on whom Iacocca had blamed the TC fiasco and who was already planning to take early retirement, a political casualty of the upheaval that had brought Bob Lutz to power. Campana had the same reservations about the Imperial as the reviewers, though naturally he never betrayed them publicly.

Campana knew as well as anyone in the industry that it was the kiss of death if dealers or customers sensed that auto executives were in the least bit equivocal about what they were touting.

Within months it was evident that luxury car buyers had little interest in the Imperial. Chrysler dealers had trouble selling the car at all, except by piling on substantial discounts. The Imperials were disposed of by the thousand to daily rental fleets, where they graced airport parking lots. The discounted, unloved Imperial set a new low in the fall of the American automobile, a sad footnote on how far Chrysler had declined from its days of glory and revival.

Robert S. Miller, Jr., known to everyone as Steve, was a bean-counter, a chief financial officer who knew little about fuel injection or aerodynamics in a company filled with passionate automotive opinions. Better than anyone at Chrysler, Miller understood how Iacocca's strategies to inflate the stock price could unravel in the event of financial miscues.

Chrysler, whose financial recovery just a few years earlier had been so robust that it barely knew what to do with its surplus cash, was heading back to the ditch it had so narrowly avoided. The automaker had spent its money on a variety of projects: $735 million on Gulfstream, hundreds of millions each on various banks and finance companies, and nearly $1.2 billion on the repurchase of Chrysler common stock from shareholders. In Miller's opinion, an argument could be made for the money spent on Gulfstream and the financial services subsidiaries. They were expected to generate income at times when automaking profits were weak; and the financial services companies, when added to Chrysler Financial Corporation's stable, were meant to reduce Chrysler Financial's reliance on automotive lending.

Shrinking Chrysler Financial's automotive loans as a proportion of total loans, Miller had figured, might serve to strengthen the financial subsidiary's shaky credit rating. Miller's job was to safeguard Chrysler's access to credit. Ratings agencies like Standard and Poor's and Moody's regarded the condition of Chrysler Financial as inexorably linked to that of Chrysler Corporation; when the automaker's financial condition was shaky, the rating assigned to financially solid Chrysler Financial reflected that shakiness.

Miller's goal was to loosen that link by diversifying Chrysler Financial's portfolio. By developing a nonautomotive loan portfolio, Chrysler Financial might be seen as a diversified company. He wanted analysts to think about Chrysler Financial as if it *were* truly independent. For Miller knew that if Chrysler's ratings dropped because of poor results, Chrysler Financial might find itself unable to qualify to sell commercial paper to investors, depriving it of access to low-interest money for loans. Banks were a theoretical source of loans but at higher interest rates; and banks weren't always willing to lend to car buyers or to finance dealer inventories of new vehicles.

In view of weakening automotive sales in 1989 and Chrysler's vulnerability to lower credit ratings, Miller was deeply alarmed by the steady outlay of cash—ordered by Iacocca—to buy Chrysler shares on the open market from investors. Iacocca had ordered the first share buyback in 1984, and steadily increased the amount Chrysler was spending for shares since then. The technique of buying back shares of a company's stock can increase share price by decreasing the total number of shares outstanding. However, the increase in share price does not necessarily indicate an increase in the public's valuation of the company. For example: if company X's capitalization is one million shares of stock selling for ten dollars each, then its total market capitalization is said to be $10 million. If company X spends $5 million cash to buy 500,000 of its own shares—*and nothing else about the company has changed*—then company X's remaining 500,000 shares theoretically should be worth twenty dollars each. In other words, the value of company X's market capitalization hasn't changed. As long as investor confidence isn't shaken and share prices rise steadily, a stock buyback may wind up to be a good deal for the company. For if the company needs capital later on, it may be able to resell shares to the public at a higher price than it bought them. Of course, another scenario is also possible: if investor confidence diminishes and the share price of company X falls, it may need to raise fresh capital by reselling shares to the public at five dollars each that it had earlier bought back for ten dollars.

Buying shares back from the public has another, delicious attraction for executives. It can be a highly effective and perfectly legal technique for self-enrichment. For a chief executive whose compensation is based on stock options, a share buyback is a way to exert upward pressure on share prices and, thus, the value of the stock options. (An executive who orders a share buyback typically defends the maneuver by arguing that his company's stock is "undervalued" and, therefore, a good way to invest the company's own cash.) A big chunk of the $40 million of Iacocca's compensation between 1986 and 1988 undoubtedly stemmed from the increased value of stock options exercised at a time when Chrysler was aggressively buying back its shares.

Steve Miller, though he had been awarded stock options as well and therefore benefited from share buybacks, sensed that Iacocca's appetite for buying in stock was a ticking time bomb. Buying out investors when times were good meant investors might not be around when times turned bad. Other Chrysler financial specialists understood this as well. Fred Zuckerman, Chrysler's treasurer, was the first and most strident to argue against using cash flow from strong automotive sales to buy back shares. Eventually, Zuckerman said, Chrysler would need the cash it was spending, and there was no guarantee that the automaker would be able to find investors.

Zuckerman, who tended to be outspoken, lobbied vigorously against buying back shares. At first he dragged his feet, buying back shares as slowly as possible. But when Iacocca ordered him to speed up, he had no choice but to comply. Zuckerman expressed his reservations to Iacocca, Miller, and others at Chrysler, as well as to bankers, members of the financial press, and his many contacts on Wall Street. Someday, he said, Chrysler might need a stronger equity base and every bit of cash it had. But in the mid-1980s, at the height of Iacocca's popularity and Chrysler's prosperity, there were few sympathetic ears for his Cassandra-like pronouncements. Under Iacocca's brilliant leadership, Chrysler had turned the corner, he was told, and there was little reason to think its future wouldn't grow brighter and brighter.

Zuckerman's self-assured, somewhat contrarian financial views went against the grain. To Iacocca's way of thinking, Zuckerman's jeremiads on share buybacks and other issues were disturbingly negative, even disloyal. He'd scream and yell at Zuckerman, and for a few weeks Zuckerman would lay low, a nonperson in Iacocca's eyes. In time, Iacocca's ire would dissipate and Zuckerman would return to his good graces. When Lee wanted to buy Gulfstream from his friend Allen Paulson and ordered that a financial case be made for it, Zuckerman made a point of ducking the assignment. He remembered how, during the bailout, Iacocca had ordered him and others to conjure a financial rationale for owning the corporate jets the government wanted sold. But Zuckerman believed there was little rationale for buying Gulfstream other than to satisfy Iacocca's personal interest in controlling a corporate jet company. Unlike other Chrysler officers, who were reluctant to see their names in print lest they provoke Iacocca, Zuckerman frequently was quoted in the financial press, often in cautionary terms that toned down Iacocca's bullish exaggerations.

Like Miller and most other Chrysler officers, Zuckerman had started his career at Ford. From Ford, where he had worked in financial management, he moved to IBM, where he spent ten years before being recruited to Chrysler by Iacocca, in 1979. At Chrysler he toiled as a subordinate to Miller and to Greenwald on the financial intricacies of the Loan Guarantee Act and the various agreements with banks and the union, rising from assistant comptroller to treasurer. A decade after leaving IBM, with Chrysler running out of money and its share of the U.S. car market dwindling, Zuckerman began to see his dire predictions take shape.

His boss, Steve Miller, was more sanguine. He believed the cars that Lutz, Castaing, and Gale wanted to develop would be easier to sell. The formation of platform teams would help transfer power from Iacocca and other executives into the hands of lower-level employees, who needed the authority to decide key isues. Miller, a brainy graduate of Stanford Business School and Harvard Law School, was viewed as the house intellectual. He shared some of the reservations of Fred Zuckerman and others about Chrysler's management and financial strength; but, unlike Zuckerman, Miller was circumspect. He often didn't agree with orders, but he performed them without argument, avoiding direct conflict whenever possible.

Miller's one provocation to Chrysler's management culture was his mentorship of Reiko McKendry, the Japanese-American woman who served on the "Honda team." Following a speech in Madison, Wisconsin, in 1984, Steve Miller found himself face-to-face with a young woman who introduced herself as an M.B.A. candidate at the University of Wisconsin and thrust a copy of her study on U.S.-Japanese trade relations into his hands. Miller didn't read the study, but he liked the woman's moxie — especially in light of Iacocca's bashing of the Japanese — and passed her name along to Chrysler recruiters. Reiko McKendry was duly contacted and offered a job. And though most M.B.A.s preferred to work on the East or West coasts rather than in Detroit, McKendry accepted. She was, however, immediately appalled by the authoritarian, haphazard nature of Chrysler's management and by Iacocca's frequent racist pronouncements. Still, she regularly discussed ideas for reform with Miller, who encouraged her to pursue her agenda. One of her proposals had been to study Honda.

The son of a lawyer from Portland, Oregon, Steve Miller grew up amid his family's timber business. After law school, he joined Ford's finance department in the late 1960s. In 1979 he was working at Ford of Venezuela when Jerry Greenwald jumped to Chrysler. Miller was soon thereafter recruited by Greenwald, and at Chrysler he became one of the central figures of the bailout, gaining a reputation for good humor and his ability to win over bankers who were reluctant to extend Chrysler credit. He had been a finance expert at Ford, dealing mostly with internal finance and the other issues facing a comptroller. Before coming to Chrysler, he rarely dealt with bankers or issues of external finance. By the time Chrysler had turned around and was making money, Miller was thoroughly experienced in external finance. He favored expanding Chrysler's presence in financial services by buying E.F. Hutton, an idea that was vetoed by Chrysler directors. He also helped negotiate the American Motors merger.

Miller's negotiating and financial skills were enhanced by a nimble intellect and a modest, self-confident manner, which he wore well on his six-foot four-inch frame. An unusual extension of Steve Miller's executive persona was his wife, Maggie. The couple formed a close marital and intellectual partnership. Unlike many corporate wives, Maggie Miller was kept abreast at all times by her husband of what was happening at Chrysler. Steve Miller regarded his wife as a shrewder judge of people than he was. She had, however, little discretion about where and when she expressed her opinion. At Chrysler's 1984 Christmas party, she could be heard talking loudly about how stupid it was to buy back shares. Zuckerman, alarmed, ran over to tell Steve Miller to please tell Maggie to pipe down before Iacocca heard her. "I've had no luck getting Maggie to shut up over the last twenty years and don't think I'll have any tonight," Miller told Zuckerman.

Maggie Miller lacked notably the customary restraint that prevents those in a corporate setting—especially corporate wives—from expressing themselves frankly. If someone flubbed an assignment, humiliated himself in the newspaper, or put on some weight, Maggie had no compunction about bringing it up. Her frankness alternately amused and horrified Chrysler officers and directors. An unrepentant smoker, she sometimes scandalized everyone present by lighting up a cigar. A lesser man might have felt threatened by his wife. Steve Miller often explained, "We're a package. You take me, you get Maggie."

Maggie Miller was also an unabashed champion of her husband's career. At cocktail-party debates about who should run Chrysler after Iacocca, she argued for Steve Miller's candidacy. Naturally Glenda Greenwald wanted her husband, Jerry, to be chairman, and Heide Lutz thought Bob should have the job. But only Maggie had the nerve to say what she was thinking in front of everyone. At a company dinner at which her husband was addressing Chrysler managers, she raised her hand during the question-and-answer period to ask her husband if he might run Chrysler differently if he were chairman. Miller, stunned by the question, put her off with humor. "The first thing I'd do," he said, "is forbid spouses from attending meetings."

Justifying the fears of Miller and Zuckerman, Chrysler found itself in 1989 wishing it had socked away some of the money it had spent to buy Gulfstream, financial services companies, and Chrysler shares—not to mention the hundreds of millions wasted on the Maserati and other ill-fated ventures. Chrysler wasn't running out of money the way it had in 1978, but it might. Fixed costs had grown rapidly since the American Motors acquisition in 1987. Bigger and bigger rebates, which cut into profit, were stimulating sales. But if sales revenues ever collapsed, costs would overwhelm the company in a matter of months. Then the company truly *would* run out of cash, and creditors might pull the plug on bank loans.

The man who ultimately sounded the alarm was no finance expert but Iacocca himself. In the summer of 1989, Iacocca summoned Ben Bidwell, his closest confidant. Bad times are coming, he told Bidwell. Sales volume was trending downward, kept alive only by rebates. The Spirit and the Acclaim were big disappointments. Chrysler's factory utilization, which had been running at more than 90 percent, was down to about 70 percent. The American Motors merger had raised costs enormously. The stock market crash of October 1987 had driven the price of Chrysler stock from the neighborhood of forty-five dollars per share to about twenty-five; with Chrysler's equity so much lower in value, it was far more difficult, and more expensive, to borrow money. This had been Zuckerman's and Miller's fear.

Ben, you've got to find ways to squeeze a billion dollars annually out of our costs, Iacocca said.

Bidwell went to tell Miller; but somehow Bidwell had misunderstood Iacocca. He thought Iacocca had said Chrysler's capital budget had to be cut by $1 billion. This was one of Detroit's standard tactics for dealing with the ups and downs of the market. When money ran low, just cut the research and development budget. Postponing new models could accomplish the required cost saving. Nothing too drastic.

Miller and Bidwell brought their ideas to Iacocca. *No,* Iacocca told them, *you didn't understand. I'm talking about a billion out of our basic costs. Our capital budget has to stay put.* Removing $1 billion from basic costs—salaries, expenses, material costs—was a whole different ballgame. Cutting capital outlays was out of the question. By this time, it was evident that no amount of plastic surgery could make the K cars attractive. Chrysler needed new vehicles, primarily a new family sedan, and that was going to cost billions.

The guy must be on something, Bidwell thought. Where would he find $1 billion worth of "fat" to trim? As Henry Ford II used to tell him, "Costs walk into your office on two legs." Offices would have to be closed, managers were going to lose their jobs, plants would shut down, workers would be out on the street. Cuts in operating costs were real, Bidwell knew, and very painful. Moreover, it wasn't going to be easy to justify firing people while Iacocca continued to live in regal style, flying everywhere by corporate jet, living in a lavish suite at the Waldorf, paying masseurs and florists and caterers from the company kitty.

Bidwell, a bit nervously, mentioned that Iacocca might start by trimming back his own corporate lifestyle.

Sure, I'll cut back some, Iacocca promised Bidwell. Bidwell had been certain that Iacocca would bite his head off at the very mention of decorating budgets and masseurs. Lee's new consort, Darrien Earle, at that very moment was helping to redecorate Chrysler's suite at the Waldorf, an undertaking that cost nearly $2 million. The suite, available to Chrysler's top officers, but used mostly by Iacocca, had been closed for removal of asbestos. Workers from Chrysler's facilities staff had redecorated the thirty-fifth-floor suite, which contained a solarium overlooking Park Avenue. But Iacocca thought it was too spartan. He missed the fireplace that had been taken out. A decorator suggested by Ms. Earle began sprucing up the suite.

Iacocca surprised Bidwell by agreeing, or at least seeming to agree, to cut back on personal frills. By Miller's reckoning, Chrysler employed roughly twenty-five people at Chrysler full-time, exclusively for the care, feeding, and personal whims of Lee Iacocca. There were pilots, flight attendants, and mechanics who essentially kept Iacocca's Gulfstream IV jet at the ready. There were security guards, masseurs, drivers, and others like Wes Small, who did little but run errands, throw parties, check menus, and make sure the chairman was happy. When Iacocca's jet landed, an entourage was always waiting. The millions of dollars these services cost were hardly enough to make a difference on the financial scale that Chrysler operated; but

the opulence sent the wrong message to Chrysler's white-collar workers. The company hardly could afford to offend its managerial rank
and file by ordering them to fire people and slash budgets while the
chairman was living like Louis XIV.

Iacocca wasn't a bit bashful about what he viewed as his entitle-
ment to his pay and perks. "I saved this goddam company," he would
mutter within earshot of his executives. The board apparently agreed:
Iacocca hadn't done it by himself, but Chrysler certainly couldn't
have survived without him. That his compensation was enormous,
even by the generous standards of Detroit, struck Iacocca as irrelevant. His role models in executive compensation were Michael Eisner
at Disney and Steve Ross at Time Warner, men who measured their
pay not in the millions but in the tens and hundreds of millions.

In late July 1989, Iacocca gathered Chrysler's top 700 executives
and managers in a concert hall in downtown Detroit and informed
them that $1 billion had to be sliced from Chrysler's costs. About
2,300 people, or 8 percent of the automaker's white-collar workforce,
had to be let go, he told them. The company was willing to pay for early
retirements and separation payments to persuade people to leave. To
encourage managers to perform the painful cuts, Iacocca announced
an incentive: managers were eligible to pledge up to 10 percent of
their paycheck as a bet that Chrysler would reach its cost-cutting
goal. If the $1 billion goal was reached, the managers received double
the amount of their pledges in cash. The top thirty-one officers of the
company were automatically entered into the program.

On October 15, 1989, Iacocca celebrated his sixty-fifth birthday.
At Ford and General Motors, as well as the most publicly held corporations, a new chief executive would already be in office and the
previous CEO would be contemplating his golden years in Naples or
Palm Springs. At Chrysler, Iacocca was studiously vague. His four-
year employment contract as chief executive was set to expire at the
end of 1991 when he would be sixty-seven. A year earlier, he had
reshuffled executives, scrapping the holding-company structure and
naming Jerry Greenwald vice chairman. Iacocca didn't want anyone
getting too comfortable in his job; yet corporate prudence dictated

that someone must be named successor, in case of unexpected trag-
edy, if for no other reason.

Thus the chairman's job became "Jerry's to lose." But the designa-
tion of Jerry Greenwald as heir apparent begged the question of when
Iacocca would actually retire. Iacocca professed to be weary of the
endless meetings, the constant travel, and the food on the banquet
circuit. One day he was attending Malcolm Forbes's birthday party in
Morocco, the next he was in California meeting dealers. As he told
reporters who asked, he didn't feel like Chrysler needed him as chief
executive but he also didn't want to foreclose his option to "stay behind to help, but in a lower profile, on the board." The message was
that no matter who was promoted to be chairman and chief executive,
Lee Iacocca intended to stick around and keep an eye on things.

Lee Iacocca's sixty-fifth birthday party was a splendid affair. The
Rainbow Room in Rockefeller Center was brightly decorated. Iacocca's pals were all invited. Gay Talese, the author, read a tribute. Video
birthday wishes from Barbra Streisand and others were played for the
audience. For Chrysler's top officers, Iacocca's birthday party was a
command performance. Maggie Miller, for one, was quite impressed.
She marched over to Iacocca. "Lee," she said, struck by the grandeur
of the celebration, "no one is ever going to call you cheap again!"

THE MAKING
OF THE L/H

THE DECISION WHO should lead the team building the L/H family sedan in early 1989 was a crucial one.

François Castaing, the newly minted vice president of engineering, needed a seasoned leader. That individual also had to understand why 900 or so engineers and planners—organized as a platform team—were expected to finish the L/H faster, cheaper, and at a higher level of quality than Chrysler engineers had ever accomplished before. In the past, when engineers and planners were "lent" from departments to projects, a new vehicle had never taken less than four and a half years from the day of executive approval to the first day of actual line assembly. A great deal of the time was wasted arguing and arbitrating; every fight that couldn't be resolved had to be decided by a senior executive. The L/H, by contrast, had to be ready in forty-two months in order to meet Iacocca's mid-1992 deadline.

Glenn Gardner, who had joined Chrysler as a young engineer trainee in 1958, had risen steadily to assignments of increasing responsibility. In 1981 he directed the minivan project. Next, he supervised development of the Dakota midsize pickup truck. In 1985 he was selected as senior Chrysler executive at the Diamond-Star joint-venture assembly plant with Mitsubishi Motors in Bloomington, Illinois. That project had afforded Gardner an opportunity to see how the Japanese operated at firsthand. It was also a chance to see how

American workers took to the process. Although Chrysler invested half the capital for the assembly plant, management of the plant was the responsibility of Mitsubishi, a company well versed in the manufacturing principles that MIT researchers had studied at Toyota: continuous improvement, consensus management, the discovery of the root causes of defects, just-in-time inventory control — skills the U.S. auto industry was only beginning to understand. Gardner realized his role at Diamond-Star was to protect Chrysler's interest but, most importantly, to learn as much as possible.

After four years at Diamond-Star, Glenn Gardner had become thoroughly converted into a disciple of Japanese manufacturing methods. While many of his contemporaries in Detroit echoed Iacocca's disparagements of Japan and their manufacturing triumphs, Gardner grew more fascinated by the day as he watched precisely how they were accomplished — in a Midwestern location, and by American workers recruited from the Illinois countryside. Gardner was struck by the patient instruction from the Japanese managers, the slow but deliberate pace of decision-making, which paid off in rapid results. Relatively junior employees received copious training and were trusted to decide wisely. As much as possible, politics and time-wasting bureaucratic rituals were eliminated from decision-making. What workers received instead were clearly stated goals and a sense of shared mission. Yet there was, Gardner came to understand, very little truly "Japanese" about Mitsubishi's accomplishments. If Chrysler and its workers kept an open mind, they could learn everything Mitsubishi knew and replicate it in Chrysler factories.

Gardner was outspoken in his fascination and admiration for the Diamond-Star joint venture. His attitude was reflected in interviews with the automotive press and in enthusiastic exchanges with engineers and managers back at Chrysler headquarters in Highland Park. Lutz and others worried that Gardner might be too pro-Japan and might alienate key managers at Chrysler and its suppliers, not to mention Iacocca himself. The alliance with Mitsubishi notwithstanding, many were convinced that Japan, or at least the Japanese auto industry, should be regarded as the enemy.

For those Chrysler managers who were upset about the forced

exit of Hal Sperlich, the appointment of Gardner was another worry.

Chrysler engineers saw the team concept as a risk. It was fine for factory workers; it was fine for the Japanese as well, but it hadn't been proven at Chrysler. Managers who had climbed the corporate ladder for twenty years weren't anxious to forgo the perquisites or the authority that came with seniority. To supervise all brake engineers was to possess power; to be the brake specialist on the L/H team, by contrast, demanded only expertise. Faced with resistance from those engineers who declined to work on the L/H team, Castaing and Lutz issued directives to managers to submit names of potential team members. Gardner met with many of them, attempting to explain the need for changing the way cars and trucks were built. He distributed dozens of copies of the Womack, Jones, and Roos book *The Machine That Changed the World.* Prospective team members were skeptical that what was customarily a fifty-four-month process could be shrunk to forty-two months. Moreover, Castaing, because of his American Motors pedigree, was becoming a target of hostility from longtime Chrysler managers who suspected that Lutz and Castaing were trying to "reform" engineering as a means of consolidating authority for themselves, the final scene in the purge of Hal Sperlich and his followers. Lutz and Castaing promised that none of the 850 employees needed for the L/H team would be forced to join. Still, many did so with less than a full heart.

In the first six months, Gardner's top assistants reported that the L/H was on schedule and could be delivered by Iacocca's mid-1992 deadline. Under Gardner's leadership, the several engineering factions demonstrated they could cooperate with marketing and finance specialists, suppliers, and one another to reach creative compromises. Many tasks were completed simultaneously rather than sequentially. In case after case, what used to take months was taking days. In addition, Gardner, Castaing, and Lutz insulated members of the team from interference and suggestions from senior executives of the sort that delayed projects and added costs. Senior executives were kept at arm's length, although they were allowed to attend periodic reviews where they could inspect models and drawings and ask questions. Gardner, with the backing of Castaing and Lutz, listened respectfully to suggestions, nodded politely, and pressed forward as originally planned.

Iacocca was notorious among engineers — as Henry Ford II had been — for "suggesting." (Ford would suggest that a car under development needed a few more inches here or a few less there — suggestions that would slow projects for months.) After one review in 1989, Iacocca sent word that he didn't care for the design of the L/H's grille. Typically, the message would have been cause for immediate redesign of the grille and whatever else was necessary to satisfy the chairman. Whatever it cost and the length of the delay mattered little; but this time Iacocca's objections, once considered, were simply put aside. Lutz was in charge of "gently" wrapping Iacocca's suggestions in a gauze of inaction, while reassuring Castaing and Gardner that they shouldn't alter any of their designs.

From a pure engineering standpoint, the L/H was expected to fulfill ambitious goals. Domestic sedans typically performed and handled in a manner inferior to Japanese sedans. One reason was that the steel panels that formed the sedan's body did not impart an overall stiffness or rigidity equal to the standards of Japanese automakers. The issue wasn't engineering expertise so much as an attitude shared by U.S. automakers that body stiffness wasn't that important, except insofar as it roughly matched that of other domestic models. Hence, while turning or braking, domestic sedans tended to sway, bend, or vibrate. Comparable Japanese models, by contrast, felt very "tight." Incredibly, U.S. manufacturers had only sporadically studied the body stiffness of Japanese models or attempted to match them. Gardner insisted that L/H body engineers use the ultrastiff Nissan Maxima body as their benchmark and encouraged them to decide on their own how to achieve that standard.

In discarding the L/H prototype built by Hal Sperlich and Jack Withrow, Lutz, Castaing, and Gale committed themselves to a new concept of engine and transmission. The engine was now aligned longitudinally, or north–south. The reasons were twofold: first, to permit Gardner's engineers to make the engine compartment smaller and apportion more space to the cabin; in the "cab forward" version, 75 percent of the vehicle's length was dedicated to passenger and cargo, compared with 65 percent in conventional vehicle bodies. Second, north–south engine alignment offered the opportunity to build a larger, rear-wheel-drive luxury version of the L/H at a later date, almost a technical impossibility with the engine mounted east–west.

Glenn Gardner was thrilled to hear in 1989 that Chrysler's "Honda team" had expressed opinions similar to those he had reached at Diamond-Star about how Chrysler had to restructure. The Kohler seminars led by Doug Anderson were further evidence that others recognized the importance of adopting a broader view about automaking and of opening minds to the ideas furnished by the Japanese. Gardner needed allies. His opinions constituted a minority view, which was provoking frustration in certain quarters.

The actual *manufacturers* — the factory rats — were always the last to be consulted. Typically, blueprints of the vehicle they were to build were delivered to them at the last possible moment. Little time was left to solve problems before the start of production. In the end, parts that didn't fit together well were tortured into submission with tools and sweat.

This time around, Castaing and Gardner made sure the factory rats were accorded full membership on the L/H team. As Toyota and others had so often demonstrated, when manufacturing engineers and plant foremen participate in vehicle design, speed and quality improve immensely. The concept was "design for manufacture," or DFM, the design of parts for ease of assembly. Autoworkers and engineers from Chrysler's Bramalea, Ontario, plant were invited to build L/H prototypes, a departure from the practice of hiring an outside source to do so. Members of the Canadian Auto Workers Union won the job of teaching assembly techniques — also a departure from tradition. Workers from Bramalea shuttled between Ontario and L/H headquarters in Auburn Hills, near Detroit, to observe pilot assembly techniques nearly two years before the start of assembly. Normally factory workers first viewed prototypes twenty-two weeks before the day the assembly line started.

Gardner and Castaing enjoyed another manufacturing advantage: shiny new Komatsu stamping presses built in Japan. The presses, designed by Mitsubishi, produced metal parts in a building adjacent to the Bramalea plant. Canadian workers learned from their Japanese supervisors how to operate the gigantic presses to produce accurate stampings. Losses from scrap metal were minimized. The metal stampings, like pieces from a jigsaw puzzle, were welded into highly accurate metal components that assured an even fit for the L/H's doors, deck lids, and windows. By involving workers, foremen, and plant managers in everyday production decisions, Lutz and Castaing hoped to elevate standards of manufacturing quality and efficiency. Workers were encouraged to search for problems to solve without fear of blame or punishment.

But fear and defensiveness had become a way of life for Chrysler plant managers and workers. Lutz wanted to replace Dick Dauch with someone who exemplified a style of management that didn't rest on a foundation of fear. Dauch was the man who effectively modernized factories in the early 1980s by frightening and intimidating those who worked for him. Dauch had worked wonders then, but his methods were fundamentally inconsistent with what Gardner had been learning at Diamond-Star.

At first, Lutz thought he might be able to get Dauch to soften his overbearing style. The two men met frequently. Lutz suggested that his colleague rent the film *Patton*. What lesson did that film teach you? Lutz asked. It showed that a great general, a great man, could be undone by a bunch of wimps, Dauch replied.

It was no use. No matter how hard Lutz tried to suggest that browbeating subordinates wasn't the way of the future at Chrysler, Dauch was unable see the life of the factory in fresh terms. From the moment Bob Lutz succeeded Hal Sperlich in 1988, tension existed between him and Dauch. Lutz had become the man in charge of vehicle design and development; Dauch, though he reported to Lutz, was in charge of the factories that produced the vehicles. In an ideal corporation, the two men should have been natural allies; but Lutz, who had been reared in Europe and was more sophisticated, saw Dauch, the Ohio farm boy, as boorish and overbearing. Dauch dressed down subordinates unmercifully in front of their peers and often lost his temper, shouting at executives and managers who didn't agree with him or who found his rah-rah manner tedious. He delivered endless sermons about Vince Lombardi and the virtues of football, which Lutz found asinine. For his part, Dauch saw in Lutz the man blocking him from promotion to higher office and a seat on the board of directors.

By 1989, Lutz decided to fire Dick Dauch. He complained about Dauch to Greenwald, who referred Lutz's complaints to Iacocca. Lee liked Dauch and credited him with vast improvements in Chrysler's plant operations. Iacocca ordered Greenwald to "counsel" Dauch, to try to influence him to calm down and relate better to the people around him. Greenwald complied. An industrial psychologist was hired to analyze and counsel Dauch, but Greenwald continued to receive complaints about Dauch from Lutz and others.

From what Lutz could see, patient counseling had a minimal effect on Dauch's style. Indeed, attempts to tone down the gridiron executive only seemed to sharpen his view that Lutz was attempting to get the better of him in their rivalry for promotion. Dauch often told people close to him that he hoped to run Chrysler one day and that Lutz was thwarting him.

In the spring of 1989, Lutz invited Dennis Pawley to Detroit for lunch, ostensibly to discuss the installation of an elevator at Lutz's home. In reality, Lutz wanted to evaluate the man. Pawley, a manufacturing executive for the Otis Elevator division of United Technologies, had gained industrywide recognition in the mid-1980s when Mazda hired him to run its automotive assembly plant in Flat Rock, Michigan. He had tangled with Mazda's Japanese managers over the extent of his authority; but he also developed an appreciation for the style of factory-floor teamwork practiced in Japanese auto plants. The word was that Pawley, who had started his career at General Motors, was eager to return to the automobile industry and Detroit. Lutz thought he had potential and offered him a job.

Dennis Pawley joined Chrysler in April 1989 to supervise the planning of future manufacturing projects. In theory, Pawley should have been working for Dauch, the chief manufacturing executive. But within a few months he was reporting directly to Lutz, bypassing Dauch altogether. He was placed in charge of the Jefferson North Assembly plant, then under construction on the east side of Detroit.

This plant, set to begin producing Jeep Grand Cherokees in early 1992, was a factory designed to demonstrate that American autoworkers, with proper training and tools, could match Japanese levels of quality and productivity.

As relations between Lutz and Dauch deteriorated, Pawley determined to stay out of the line of fire. He was Lutz's subordinate but he could not afford to become an enemy of Dauch. Lutz was shrewd enough to understand that Pawley's future depended on his ability to avoid conflict between himself and Dauch. In some ways, Pawley could be as political and autocratic as any old-style manufacturing manager when the situation called for swift, decisive behavior, but he also comprehended the broader trends that were inevitable for factories of the future. Like Gardner, he had read *The Machine That Changed the World* many times and made subordinates read it. When Iacocca asked Pawley what he thought of the book, he told Chrysler's chairman it was the correct analysis of the U.S. auto industry's problems.

In April 1991, Dick Dauch abruptly "quit," supposedly because Iacocca would be leaving soon. As he told the *Wall Street Journal*, "I have immense respect for Mr. Iacocca, but there is an end of an era coming and it is time to move on." In fact, Dauch's resignation was a face-saving gesture. Iacocca, weary of the friction between Dauch and the rest of the company, had decided he had to let him go. Dauch was old-school. Iacocca had promised at the officers meeting two years earlier that "intimidation was out and consensus was in." Those who couldn't change had to leave.

The day Dauch's resignation was announced, Iacocca summoned Dennis Pawley to his office to promote him to vice president of manufacturing. On that day, Chrysler announced that most of Acustar, its parts-making operation, was being absorbed back into the overall manufacturing operations. Acustar had been created three years earlier, in part to remove parts-making operations from Dauch's purview; still, it had taken three years since Acustar's formation for Lutz and others to pry Dick Dauch and his influence away from manufacturing operations. To everyone who had witnessed Dauch's deprecations of subordinates or whoever felt his iron handshake, his departure spoke poignantly about Chrysler's new willingness to dream ambitious dreams for its factories and its workers.

A chief executive, even one as powerfully independent as Lee Iacocca, was supposed to run a corporation under the watchful eyes of a board of directors. According to the laws under which most corporations are formed, directors represent and are elected by shareholders; they are the ones who appoint and evaluate top management. As a practical matter, however, directors serve mostly as advisers. By and large, they are nominated by chief executives, and in most cases they exercise only loose supervision of top managers.

Due to the sea change in the U.S. auto industry during the mid-1980s, the issue of whether directors of public corporations must exert more forceful supervision moved swiftly to the forefront of the debate about the nation's industrial competitiveness. For decades, directors had served as little more than sounding boards. Often corporate boards acted as no more than rubber stamps for the chief executive, who nominated directors that could be depended upon to offer counsel and little criticism. Except for a sprinkling of token minorities and academics, directors were mostly fellow male, white, middle-aged chief executives with similar attitudes and interests. Many are retired and looking for a way to spend their time productively.

When corporations performed profitably and paid dividends to stockholders, the fiction that shareholders selected the directors as their representatives and that directors played an authentic oversight role was largely ignored. But when institutions such as General Motors and IBM floundered because of managerial incompetence, obvious questions arose: Where were the directors? Why weren't they supervising management? Why weren't shareholders holding directors accountable? What did directors truly understand about the business of the corporation they were supervising?

Ross Perot complained publicly in 1986 about GM directors' failure to act on the corporation's behalf, ridiculing the board as a "pet rock." He recognized the Japanese competitive threat and was astonished that the automaker didn't respond more forcefully. Perot was

angry with what he saw as Roger Smith's failure to live up to GM's
merger agreement with Electronic Data Services, the company Perot
had founded. The GM directors subsequently voted to buy back Perot's stake in General Motors for $700 million, in return for his
agreement to resign from the board. They were happy to be rid of
Perot, whom they regarded as a gadfly. Smith, GM's chairman and
Perot's nemesis, served for another four years after Perot's departure.
During that time, GM's directors fretted among themselves about
Smith and the direction he was taking the automaker. They agonized
over how to control Smith without compromising his authority in the
eyes of the corporation. In its worry over how not to be too strong, the
board was rendered weak. Smith actually convinced directors to alter
corporate retirement rules near the end of his term in order to enrich
himself with a million-dollar-a-year pension—a curious reward for
the market share that GM had lost to the Japanese during his tenure.

During and after Smith's years at GM, the corporation's nonemployee directors received counsel from Ira B. Millstein, a lawyer and
specialist in corporate governance. Millstein urged the directors to
probe deeper into GM's affairs, ask tougher questions of the executives—including Robert C. Stempel, Roger Smith's successor—and
assert their legal responsibility as supervisors and evaluators of top
management. Millstein and others realized that unless directors accepted greater responsibility for managerial performance, state employee pension funds and other large shareholders were liable to
revolt, causing financial and legal chaos. As it was, corporate raiders
were attracting the votes of shareholders across the land who were fed
up with ossified management.

Slowly but surely, GM directors toughened their oversight of the
corporation. They shot down Stempel's plan to pack the GM board
with his own nominees, demoted Stempel's hand-picked number two,
Lloyd Reuss, and, in late 1992, forced Stempel's resignation in favor
of Jack Smith, an executive the board believed would be a more effective leader.

Sympathetic vibrations of corporate governance issues reverber-
ated across the Motor City. Directors unhappy with the succession

plan of Don Petersen, Ford's chairman, and concerned that Petersen was losing his grip, encouraged him to take early retirement and promoted Red Poling to take his place. Board actions like those taken at Ford and General Motors would have been unthinkable only a few years earlier. Boards of directors and their lawyers began to watch one another. Directors were more concerned than at any time in their business careers about responsibility and liability. They also considered their reputations and the trashing they were getting at the hands of mass media as a result of proud corporations like GM falling into disarray. The U.S. auto industry, the backbone of American manufacturing, was sinking under the weight of outmoded management while directors tried to determine how much activism was appropriate. Painful restructuring was vital, but without firm prodding, the executives seemed disinclined to act.

Of all Detroit's chief executives, Lee Iacocca, a figure who inspired awe in the business community, enjoyed by far the longest leash. Hence, Chrysler's board was inclined to let him exercise his own judgment. Iacocca's arrival had been a key factor in the automaker's survival against heavy odds in the early 1980s. As corporate pitchman, Iacocca played a central, personal role in the selling of Chrysler cars. Directors felt little reason to question the time and effort Iacocca devoted to the Statue of Liberty, since its patriotic theme blended directly with the selling of Chrysler vehicles. Lee Iacocca was more than a chief executive, he was a one-man marketing extravaganza whose persona was inextricably entwined with that of Chrysler.

Of the fifteen or so outside members of Chrysler's board, Iacocca preferred to confer seriously with only a few. In particular, he liked to bounce ideas and proposals off J. Paul Sticht, a former chairman of RJ Reynolds, who informally surveyed other board members and solicited comment, sometimes acting as "lead board member." Jean de Grandpré, chairman of Bell Canada, and Bob Lanigan, chairman of Owens-Illinois, sometimes took the lead. As such, they were able to tell Iacocca which of his ideas was likely to fly and which was not, thus allowing Iacocca to avoid disagreements or confrontation with the board. In this sense, Chrysler's board was not unusual; many boards tend to coalesce around a single director or a small group, a kitchen cabinet, who act as a sounding board for the chief executive. This mechanism is a strength but also a weakness, for the chief executive acts as a filter for information, naturally expressing his own opinions on issues, opinions which may not reflect other important views.

Toward the end of Iacocca's term of office, the relationship between CEO and kitchen cabinet broke down. In his early years, Chrysler directors had no reason to argue with success. They agreed to Lee's forays into nonautomotive diversifications, approved the nearly $2 billion share buyback, and paid him more than any auto executive in history. They acquiesced to unorthodox moves, like nominating Doug Fraser, president of the United Auto Workers, to the board. Fraser's nomination was innovative since nowhere, except in Germany, did a leader of a trade union serve as a corporate director. Fraser's membership on the board begot constructive ideas about improving relationships with workers and the union. Also, Fraser's seat on the board enabled him more easily to persuade the union rank and file to grant Chrysler financial concessions during the bailout.

By the late 1980s, however, Chrysler's performance had begun to slip. Fraser had retired as president of the UAW and was replaced by Owen Bieber. Iacocca hadn't wanted to appoint Bieber a director, fearing he wouldn't blend in as well as Fraser had. Nevertheless, he appointed Bieber rather than risk alienating the union. Almost immediately, Iacocca came to believe he had made a mistake, that Bieber's presence spoiled the chemistry he had enjoyed with the board. To avoid including Bieber in conversations, Iacocca began conferring regularly, and privately, with a small number of board members.

Bieber, a hulking, skilled tradesman from western Michigan, was temperamentally quite different from Fraser, who, though an authentic son of the working class, could hobnob with chief executives at an expensive restaurant without feeling he was betraying his union brothers. Bieber, on the other hand, felt quite uncomfortable with the trappings of wealth and privilege. His service as a director started at a time when money was pouring into Chrysler's coffers, so, unlike Fraser, who had served during a period of austerity, Bieber had little reason for sympathy when Iacocca talked about containing wage and benefit increases. He took a tough line on diversification into nonautomotive businesses and on what he regarded as excessive salaries and benefits for executives and directors. He donated his own director's salary of $33,000 a year to his alma mater, Grand Valley State University.

The UAW president was not shy about voicing his views in long, windy speeches that Iacocca found boring. Like the Honda team and such progressive managers as Glenn Gardner, Bieber urged a version of greater worker involvement; the difference was that his ideas were set in the context of adversarial relations between automaker and union. Indeed, in many board deliberations he cast the only dissenting vote, an annoyance to Iacocca. It was, then, specifically because of Bieber that Lee kept more limited counsel than he might have, and Bieber came to suspect that important issues were being decided by the kitchen cabinet prior to board meetings. Agenda items were discussed ahead of time, approved prior to meetings by board committees, and then put to a ceremonial vote at meetings of the full board. Bieber didn't like what was happening, but he also didn't believe he should withdraw "labor's voice" from the boardroom by quitting.

In the spring of 1991, Iacocca decided to remove Bieber and four other directors from the board. Chrysler was in worse financial shape than at any time since 1980. Iacocca's cover story for the purge was that reducing the size of the board saved money, although the savings was insignificant in comparison to the $3 billion in total costs that Chrysler was attempting to cut. To be sure, the purge was aimed squarely at Bieber, whose presence at board meetings Iacocca and his confidants detested. (The other four directors who were told they were not going to be renominated at the annual meeting were Anthony J. A. Bryan, retired chairman of Copperweld; Ronald H. Grierson, vice chairman of General Electric of Britain; Juanita M. Kreps, former secretary of commerce; and J. Paul Sticht, who had stayed on for a year beyond the normal retirement age of seventy-two for directors.)

Slicing the number of outside directors from fifteen to ten made reaching a consensus easier. With each passing month, the need to find a successor to Iacocca became more pressing for the outside directors. Bankers were reluctant to renew the $2.6 billion of credit without certain assurances. They wanted to know if new cars were under development, who was being groomed for chairman, and the likely impact of Iacocca's retirement on the company. Monumental decisions would have to be made by directors and managers if Chrysler hoped to avoid a collision with financial reality. A smaller board was arguably a mightier board—which was what it might take to win in a clash with Iacocca.

CHANGING OF THE GUARD

*As I clean out my desk drawers in Highland Park, I may get
a little misty-eyed but I doubt I'm going to miss the perquisites of
power. Getting addicted to all the glory can leave you with a terrible downer
if you're not careful. Being waited on and having your royal ass kissed
for so long by so many people can leave you a little helpless.*
—LEE IACOCCA, *TALKING STRAIGHT*

THE FUTURE OF the American automobile business looked decidedly gloomy from Lee Iacocca's vantage point as the new decade began.

Chrysler's stock was hovering in the neighborhood of fifteen dollars a share (a third of its 1987 value), automotive sales were weak, and bond-rating agencies had downgraded the debt securities of Chrysler Financial Corporation, threatening the subsidiary's ability to raise money for the financing of retail sales and dealer inventories. Despite Iacocca's pleas to the public, the government, to whomever would listen, most lawmakers weren't inclined to burden Japanese automakers with tariffs, quotas, and local-content regulations, much less bar the entry of imports.

Absent government help, Iacocca believed an ideal solution to Chrysler's malaise was a mega-merger with a financially stronger company, one that understood the auto industry and fit nicely with Chrysler's operations. This description fit Fiat perfectly. The Italian automaker's stock was strong, and it built and sold its vehicles in Europe, where Chrysler was not a competitor. Iacocca, who regularly disparaged the WASP business elite in this country, was in awe of the aristocratic Agnelli family that controlled the Fiat industrial empire and of its patriarch, Giovanni Agnelli. In addition to helping Chrysler, a merger between the two companies would elevate Iacocca to a position in European industry higher than the one that Henry Ford had occupied.

The merger talks started quite modestly. Joe Cappy and Mike Hammes, two Chrysler executives responsible for international operations, met in April with Fiat officials in Turin as part of a European trip to seek out joint ventures for Chrysler. Hammes planned to leave Chrysler soon and was showing Cappy, the former president of American Motors, the ropes. Chrysler had once considered buying a small engine from Fiat. But when Chrysler Financial securities were downgraded in early 1990, the goal became more ambitious: Iacocca wanted to find out if Fiat or others might be willing to invest in Chrysler to help prop up its sagging balance sheet.

During the conversations in Turin, Hammes hinted to Fiat officials that Iacocca wasn't opposed to the idea of a full-fledged merger. A merger, Hammes said, was controversial among some top Chrysler executives who were nervous about losing their jobs; but Iacocca believed that an extended combination of the two empires might be mutually beneficial. At a subsequent meeting, Fiat officials told Hammes and Cappy, to their surprise and pleasure, that Agnelli was quite interested in any "industrial opportunity" between the two automakers. A mating dance had begun. The Chrysler executives drove to Iacocca's villa in Siena. Iacocca met them at a small hotel and was thrilled to hear of Agnelli's interest. All through April and May, a series of meetings took place in Monte Carlo, Turin, and Rome; the meetings included Iacocca, Agnelli, and several lower-ranking officials of the automakers, plus investment bankers from Salomon Brothers and the Blackstone group for Chrysler and Lazard Frères for Fiat. The Americans were struck by the heavy security surrounding the business executives, apparently a result of recent kidnappings and murders in Europe by left-wing terrorists. Security men with shot-guns constantly surrounded Iacocca's villa, and his car was accompa-nied by police cruisers filled with Uzi-toting guards.

Superficially, at least, the two companies seemed made for each other. Fiat's Lancia and Alfa brands were floundering in Europe. Fiat wanted to bundle them with Chrysler's Jeep brand and create one strong network of dealers selling all three. Fiat also wanted to sell Chrysler's minivan through its own network. Both automakers thought they might gain efficiencies of scale by purchasing parts jointly, and both needed a new four-cylinder engine, which they could develop jointly. Together, they might save perhaps as much as $3 billion.

Chrysler executives were a bit unnerved, however, to learn that Fiat wasn't building a single vehicle that they considered competitive in the United States; and none was on the drawing board. With much expense and effort, Fiat's Tipo small car might be adapted for the American market. Bob Lutz, who joined the negotiations along with Steve Miller, concluded that Fiat had little or nothing to offer Chrysler in the way of automaking expertise or ideas. He also believed that Chrysler should design its own small car rather than import the Tipo or build it under license. The L/H project, moreover, was going to prove that Chrysler could design and build cars that competed with the Japanese. The essence of any automaker was its ability *to make cars*. Lutz watched, listened, and asked questions, but he was increas-ingly opposed to a merger with the Italian automaker. To him, selling out to Fiat was a defeatist strategy.

Still, Fiat possessed something that attracted Iacocca greatly: fresh capital. Chrysler's bankers worked on marriage scenarios that would give Fiat the vehicles it wanted and Chrysler the capital it needed. One of the bankers proposed a global company akin to Royal Dutch Shell, with separate boards of directors but one overall man-agement, based perhaps in London. Iacocca began to fantasize him-self the globe-trotting monarch of an automotive empire. "If you can put together a deal like that," he told the bankers eagerly, "I'll stay for another five years."

Agnelli and the Fiat executives worried about possible social and economic liabilities, including Chrysler's pension fund which was under-funded by $3 billion. The headlines about closed plants and laid-off workers in the United States conjured images of enormous financial outlays, about which European employers were particularly careful because their own legal obligations to workers were so extensive. Fiat executives were also skeptical that Chrysler's predictions that the L/H, the Jeep Grand Cherokee, and a new small car could be built as quickly and for as little cost as projected.

Chrysler thought Fiat should buy its stock at about forty dollars per share—an investment of more than $8 billion—nearly a 300 percent premium over the market price. Fiat was prepared to invest $1 billion, $2 billion at most. In the meantime, the specter of war in the Middle East suddenly loomed in the fall when Iraq invaded Kuwait and the United States threatened to intervene. As nervous holders of Fiat shares turned jittery, the price of Fiat shares began falling. By the end of October, the price disparity, in addition to a host of other negative factors, led Agnelli and Iacocca to break off talks and announce that there would be no deal.

On November 2, 1990, the automakers held a joint press conference in New York to say they failed to reach an agreement, citing "poor economic conditions." To make sure everyone knew just who would have been buying whom, Agnelli pointed out to reporters that Fiat would have had a majority interest in the combined corporation.

World events—and some unrealistic expectations on Agnelli's checkbook—had conspired to prevent Iacocca from becoming the automaker to the Italians.

Jerry Greenwald's decision in 1990 to leave Chrysler in order to attempt the restructuring of United Air Lines upset a delicate balance of power among Chrysler executives.

Of the three remaining senior executives—Bob Lutz, Steve Miller, and Ben Bidwell—only Lutz and Miller were potential candidates to replace Greenwald as number two and heir apparent. A year earlier, Bidwell had asked Iacocca for clearance to retire and was turned down—he needed Iacocca's permission to qualify for a full pension. But in the spring of 1990, when the sixty-three-year-old

Bidwell was ordered by his cardiologist to slow down, Iacocca granted his request.

Although not a candidate, Ben Bidwell remained a factor in the executive succession. The truce that Lutz and Bidwell had forged in 1989 when Iacocca appointed them co-presidents had held. Instead of fighting over turf and additional responsibility, the two executives simply divided the jobs to be done. The arrangement didn't spring from friendship but from a concern that the floundering automaker desperately needed some cooperation among its executives. Bidwell knew he was too old to be considered seriously for Iacocca's job, and, anyway, he didn't relish the idea of a nasty succession fight. The board might turn to him to run Chrysler in an interim capacity, but no more than that.

Miller and Lutz also declared a truce. Each wanted very much to succeed Lee Iacocca, but neither was interested in fighting the other for the job. Miller was, by nature, mild-mannered, and Lutz, already regarding himself as the leading candidate, didn't feel threatened by Miller. In a series of interviews, Miller and Lutz reported to the automotive press that each would be happy to serve under the other. If Miller got Iacocca's job, Lutz was willing to be number two, and vice versa.

Chrysler could scarcely afford high-level dissension. In June 1990, Standard and Poor's dropped Chrysler's debt ratings to BBB—, only one category higher than subinvestment grade, or "junk" rating. The lower grade meant that Chrysler was perilously close to the trigger point at which banks, insurance companies, and other financial institutions would be prohibited by their internal guidelines from owning Chrysler bonds and commercial paper. Ratings agencies were worried about weak earnings, the exodus of executives—particularly Greenwald—and the stiffer competition that Chrysler vehicles were encountering. The Jeep Cherokee, one of Chrysler's most profitable models, had been the sport-utility vehicle of choice until Ford introduced its Explorer and GM brought out a four-door version of its Blazer, both at lower prices. Chrysler countered by offering some optional equipment as standard, the effect being to drop its own price

by a few thousand dollars. Still, the Explorer was blowing the Cherokee away in one-on-one comparisons.

The reasons for the financial squeeze and the weakness of
Chrysler sales were twofold: a weak economy and the failure to develop new models several years earlier. Senior executives believed
that by 1993, if all went according to schedule, Chrysler would have
the new models it needed to compete. A restyled minivan was under
development; a new Jeep Grand Cherokee was slated for introduction
in early 1992; and the L/H sedans were scheduled for mid-1992,
and a new pickup truck shortly thereafter. They were confident these
models would be popular and profitable. But new models wouldn't
matter if Chrysler went out of business in the meantime.

In Bob Lutz's opinion, Chrysler's troubles were the fault of Iacocca. By imposing his taste and sticking too long with the K cars, by
neglecting to expedite new-model development five or six years ago,
by attempting to meddle with the platform teams of engineers that
Castaing had formed, Iacocca was threatening the new models that
Lutz was convinced could save the automaker. Trying to sell Chrysler
to Fiat had been a last straw, a ploy supposedly for Chrysler's benefit
but which, in truth, had more to do with Iacocca's personal power and
aggrandizement.

Lutz became more and more vocal in his criticism of Iacocca. He
had never been terribly discreet. Now he put caution aside as he
vented his anger and frustration with his boss. "Can you believe what
that old geezer said today?" Lutz thundered at a meeting of Castaing
and other engineering executives in his office. Lutz frequently held
court in the fifth-floor suite he shared with Bidwell. As his colleagues
snickered nervously, Lutz ridiculed Iacocca's age, his affection for
chrome trim and vinyl roofs, his habit of turning to rebates and low-
interest loans to attract customers.

For an entire year, Iacocca had been dragging his heels on Lutz
and Castaing's pet project, the Viper. Sure it looks good, but show me
a business plan that justifies the expenditure, he kept demanding.
The Viper, powered by a ten-cylinder engine and able to accelerate
to sixty miles per hour in less than four seconds, was displayed as a

one-of-a-kind concept model at the 1989 Detroit Auto Show. Almost unexpectedly, it turned into an instant hit with automotive enthusiasts. Although it would require a $70 million investment to set up a small factory, Lutz and Castaing wanted to build a few thousand Vipers annually for the small number of enthusiasts who were ready to spend $40,000 to $50,000 to own a very fast, highly impractical machine. The point wasn't profit, anyway, but promotional value. The car magazines were sure to feature it on their covers. Jay Leno was a lock to buy one and have his picture taken in it. Iacocca had spent *five times as much* on his pet car, the ill-fated Maserati TC. Maybe he was just jealous; maybe he didn't want two subordinates to show him up with a car that had every indication of being a winner.

The Viper was one more instance of friction between Iacocca and Lutz. Lutz had ambitious plans, and Iacocca seemed always to be squelching them. In return, Lutz, who loved to mimic people he didn't like, mocked and derided Iacocca whenever he had a chance. His performances were even known to spill over to the outer offices, where they were overheard by people who admired Iacocca. Shortly after Greenwald's resignation, Lutz began using Bidwell as a sounding board. Bidwell listened to Lutz rail against Iacocca, trying to remain equivocal. He understood that there was substance to Lutz's complaints that Iacocca had outlived his usefulness at Chrysler, that he had stayed on as chief executive a few years too long, spent corporate money too freely, paid himself too handsomely, wielded power too arrogantly, and had used Chrysler as a means of fulfilling ludicrous, personal fantasies like the global automotive empire.

But Bidwell also felt fierce loyalty to Iacocca and admired his tenacity. Lee had never retreated at Ford even when Henry Ford II held all the aces. To be sure, he had warts, but he refused to back down during tough times. Without his presence and powers of persuasion, the government never would have guaranteed the loans that saved Chrysler. He could have bowed out gracefully starting in 1988 when business began turning sour, and he had all the money he needed. Instead, he undertook to cut operating costs earlier than all his competitors. He searched for alliances and financing, exhorted the troops. All right, so he'd gotten a little sloppy during Chrysler's prosperity, but in adversity he was as great on his own field of combat as Winston Churchill had been on his.

If Bob Lutz hoped to win an ally by voicing his disdain for Iacocca in front of Ben Bidwell, he erred. Bidwell appreciated Lutz's eye for fine cars, his ability to marshall engineering talent, and his instincts as a businessman. But his unrelenting, uncontrolled bashing of Iacocca raised questions in Bidwell's mind about Lutz's judgment as an executive. Over the years, he had listened to Lutz demean Hal Sperlich's automaking skills and belittle Jerry Greenwald as a beancounter. Now Iacocca had become the target. He wondered if perhaps what some oldtimers at Ford had always maintained was true, that Bob Lutz was fundamentally interested in the welfare of Bob Lutz and didn't care who got between him and his next promotion.

With only a few months to go before his own retirement at the end of 1990, Bidwell decided not to tell Iacocca what Lutz was saying, although Lee Iacocca was already hearing from other sources what Lutz was saying about him. One of Lutz's former secretaries, who had since transferred to another office, was so disturbed by Lutz's anti-Iacocca diatribes that she passed word of them on to Iacocca via intermediaries. Iacocca, wanting to hear what was said with his own ears, arranged to interview the secretary, discreetly, at a private location away from Chrysler headquarters.

Iacocca must have felt at that moment very much as Henry Ford II did years earlier when he learned of the rumors that Lee Iacocca was plotting against him. Did Lutz dare think he could push Iacocca aside and take Chrysler away from him? The reports of Lutz's disloyal statements drew Iacocca's attention to other evidence that Lutz and his cohorts meant to undermine the chairman and force him out of Chrysler. Ever since Lutz's formation of the L/H team and his reorganization of engineering, Iacocca had been receiving letters, some signed, some anonymous, from Chrysler managers and engineers. The mail may or may not have been an orchestrated campaign, but all the letters followed a basic theme: *The platform teams forced on us by Bob Lutz and François Castaing are causing massive waste and confusion.*

Doron P. Levin

The L/H is not going to be ready. The new Jeep won't be ready. The new pickup won't be ready. If you don't do something, the company you saved will go down the drain.

In short order, Lutz was summoned to Iacocca's office. The chairman demanded to know the meaning of the letters. (He was too clever to reveal that he had heard about the insults Lutz was spreading behind his back.) Dissension among engineers and planners was a danger that could not be overlooked. Was Lutz sure the platform teams were capable of bringing out the new models on schedule? After all, longtime Chrysler engineering executives were complaining of wasted effort, lost time, and a lack of direction.

Iacocca demanded that Lutz fire Castaing and get a new vice president of engineering. Once more the scene and the dialogue bore eerie similarities to the drama in the mid-1970s when Henry Ford II, feeling his authority threatened, ordered Iacocca to fire Hal Sperlich. This time, Iacocca had risen from the role of ambitious pretender to the throne; now *he* was the aging, paranoid king. Lutz assured Iacocca that the engineering reorganization under Castaing's command was proceeding well, and that the L/H was on schedule and under budget. Some executives and managers were bitching, Lutz told his boss, because they didn't like power and authority being taken away from them. Platforms were, by their very nature, teams, not hierarchies. Specialists from various disciplines were being compelled to cooperate with one another, some for the first time in their professional careers. The experience called for a fundamental reorientation of attitudes and values. A designer had to think not only about the shape and characteristics of a brake pedal but also about how a worker on the line would assemble it. Purchasing managers had to build long-term relationships with outside suppliers instead of pitting one against the other in order to knock a few cents off the price of a part. Department heads had to relinquish authority to subordinates, who, in turn, had to learn to make decisions about what was best for the vehicle and the customer—not what made the boss look good.

Most importantly, the notions of pay, perquisites, and career had shifted. No longer was one supposed to rise to a better job or higher pay grade by hitching his or her wagon to the right boss's star. Engi-

neers and managers were evaluated by the unselfishness of their contribution to the overall project.

In theory, Iacocca understood and agreed with the principles of
teamwork, but in reality they were antithetical to his experience. He
had complimented the young Chrysler managers who produced the
Honda study. He had participated in seminars aimed at promoting
greater cooperation among executives. But in practice, Iacocca was
uncomfortable with the inherent challenges that this "cultural"
change posed to everything he had ever believed in his forty-year
career. He hated being closed out of decisions he would have made
in bygone years. Worse still, Lutz seemed determined to make him
look like a senile old fool; no doubt Lutz intended to rewrite history
after he left. All of Iacocca's accomplishments would be forgotten or
made to appear shabby, or even dishonest. Well, he would not let that
happen. Bob Lutz was not taking his company from him. Iacocca was
sure of that. There were few certainties in life, but this was one of
them.

Ever the warrior, Lee Iacocca determined that until new models
were ready he must use his personal credibility to flog what Chrysler
had in its showrooms. In 1990, the preponderance of market research
showed that buyers were flocking to Japanese brands over Chrysler in
ever increasing numbers, particularly on the coasts and throughout
the Sunbelt; moreover, buyers were more satisfied with the treatment
they were getting from the dealers of Japanese brands than from
Chrysler dealers. J. D. Power and Associates, *Consumer Reports,* and
private research all confirmed these broad trends. Chrysler managers
learned by studying surveys of customers who bought Dodge and
Plymouth models imported from Japan—built by Mitsubishi—that
Japanese-built vehicles held a substantive advantage in quality, du-
rability, and "softer" factors like ease of handling and the ergonomic
friendliness of buttons, knobs, and dials.

Since Chrysler was unable to do much about the vehicles it was
currently producing, the automaker somehow had to convince consumers that its vehicles were competitive—not inferior—to Hondas,
Toyotas, and other Japanese brands. To accomplish this, Chrysler

needed research findings that "proved" its contention. To create such research was an effort but not impossible. Depending on how the sample was chosen and how the items on the questionnaires were worded, a skilled researcher might be able to demonstrate that *for a certain group of people* Chrysler's cars stacked up well against Honda's. A test like this would never be accepted by sophisticated marketers, but it could be useful in advertising. The purpose of Pepsi Cola's controversial taste tests ("the Pepsi challenge") that "proved" people actually preferred Pepsi over Coke were never meant to stand up as valid scientific research, only to sell more Pepsi. By carefully selecting test subjects and trying a variety of research protocols, it is not terribly difficult to produce isolated results that show consumers preferring one brand over another. Citing results from driving tests held in California by the U.S. Testing Company of Wheeling, Indiana, Chrysler declared in a series of 1990 advertisements that 83 percent of drivers who said they were considering a foreign car preferred driving a Chrysler over a Honda Accord. In similar tests, 76 percent preferred driving the Dodge Shadow and Plymouth Sundance over the Honda Civic.

Most car buyers were unlikely to know that the U.S. Testing Company was a relatively obscure research company whose Chrysler test data was of questionable validity or relevance. The point of the tests was simply to give Lee Iacocca ammunition to use in company advertising: "I say our cars are every bit as good as Japanese. You say, 'I'm from Missouri, prove it.' . . . You wanted proof. You got proof."

Throughout the spring of 1990, Iacocca barnstormed six large U.S. cities, starting in Washington, to confront what he called "the myth of Japanese superiority." Surrounded by models of the Viper, the Mitsubishi-produced Stealth, and the minivan, Iacocca declaimed from the podium and pressed the flesh, flanked by important invited guests from the fields of law, business, medicine, and government. Anti-Japanese quotes, some with an undertone of derogatory racial imagery, were fed to the local press. And to fan the xenophobia of some consumers, Iacocca repeated, every chance he got, that Japanese automakers were wearing a "Teflon kimono" that protected them from harsh judgments for their unfair economic behavior. Sounding very much like Jimmy Carter complaining about national malaise, Iacocca attacked what he called "a national inferiority complex" that prevented consumers from realizing the positive characteristics of Chryslers and other American cars.

Image, Iacocca knew, equaled reality when it came to selling consumer products. Shroud Chrysler in an aura of defeat and doom and suddenly its cars will have the smell of death about them; prove that Chrysler cars have been underappreciated and unfairly judged and they rebound in the consumer's estimation.

Predictably, Iacocca's publicity apparatus blamed the press for putting a negative spin on Chrysler's condition. Reportage of Chrysler was derived from a variety of standard sources: quarterly financial reports, ten-day reports on the industry's retail sales, evaluations by consumer-testing companies, reviews by automotive publications, and personal reporting. Chrysler executives complained that ten-day sales reports from manufacturers tended to exaggerate Chrysler's losses in market share because they focused on the figures for cars rather than light trucks, the category that included Chrysler's minivans. In a speech to the *Automotive News* World Congress in Detroit, Ben Bidwell lambasted the press as "congenital sickos" for daring to compare Chrysler's current sales woes to its fight for survival during the bailout.

From a financial standpoint, several unprofitable quarters had brought Chrysler not to the brink, but rather, as some analysts wisecracked, to "the brink of the brink." Much as an army marches on its stomach, an automaker marches on its cash flow. Cash flow, in turn, depends on the confidence and forebearance of lenders. Iacocca finally felt constrained in 1990 to explain directly why the miracle company that had risen from the ashes had fallen onto hard times. Blaming Japan and the press had been empty exercises, and they persuaded few very hard-headed investors or lenders.

Late in the summer, Iacocca conceded in newspaper and magazine interviews that diversifying into nonautomotive companies had been a giant mistake. To underscore this judgment, Chrysler had agreed to sell Gulfstream Aerospace Corporation, the maker of corporate jets that it had owned for four years, to a leveraged buyout firm,

Forstmann, Little and Company. Diversification hadn't been a mistake because it diverted money that Chrysler needed for vehicle development, Iacocca insisted, but because wheeling and dealing diverted the attention of managers from the central mission of developing new vehicles. Iacocca now admitted that the notion of creating a holding company was grandiose and unnecessary; and he took the rap for his beloved but failed Maserati TC. But he drew the line at questions about the Statue of Liberty, his Italian villa, his brief marriage to Peggy Johnson (and the messy divorce), and the distractions of celebrity—the celebrity that, in its raw form, attracted buyers to dealer showrooms.

When asked, Iacocca glossed over the Honda team and the Kohler meetings, denying that they were significant. He knew they were crucial, but he wasn't prepared to admit publicly that his style of management needed an overhaul. Chrysler and other manufacturers had experimented with executive self-help seminars. Seminars meant to improve communication were nothing more than Band-Aids, a device to ease pressure. Of the Kohler meetings, Iacocca said only that he had begun conducting "town hall" meetings of 175 randomly selected employees who were urged to ask questions and express frustrations.

"I talk for five minutes and then they get an hour and a half to knock my head in," was Iacocca's description of the meetings to the *Wall Street Journal.* As for the inevitable questions about his retirement, Iacocca said he was watching Lutz and Miller carefully and planned to make a decision sometime in the next year.

Any hope that auto sales might perk up, improving the outlook of Chrysler and other U.S. automakers, vanished in the fall of 1990. With a massive buildup of arms in the Persian Gulf, George Bush signaled the administration's intent to oust Saddam Hussein from Kuwait. As the nation girded for military action, only those consumers who absolutely needed a new vehicle seemed to be in the market. Bob Stempel at General Motors publicly lamented his rotten luck at being named GM chairman during one of the worst industry slumps in history.

The price of Chrysler's stock, which had been hovering at about fifteen dollars a share during the summer, drifted toward ten dollars a share. For less than $3 billion, a wealthy tycoon could own the automaker; and there were some who went as far as to have their bankers run the numbers. On the afternoon of December 14, Chrysler issued a terse announcement that Kirk Kerkorian, a reclusive California investor, had been buying Chrysler common stock and had amassed a 9.8 percent stake, worth about $277 million. Managers in Chrysler's treasury office had been monitoring the unusually large purchases of shares by Bear Stearns brokers. That day, Iacocca had swung into action, convening a meeting of the board to enact by-law provisions that caused a purchase of 10 percent of the company's stock to trigger a set of "poison pill" provisions that would make an unfriendly takeover of Chrysler impossible.

The provisions were a safeguard, but also a warning shot across Kerkorian's bow. No one had a clue about his agenda. No one at Chrysler had ever met him except for Iacocca, who had been introduced to him once at a social event. Chrysler's lawyers and investment advisers from the Blackstone Group and Salomon Brothers advised caution. Kerkorian's reputation was mixed. The seventy-three-year-old investor was experienced and shrewd, having at one time owned sizable stakes in Western Airlines, M-G-M, and Columbia Pictures. But unless Kerkorian was restricted legally, the bankers warned, he was liable to meddle in Chrysler's business affairs and distract management. With its credit ratings weak, Chrysler was in the midst of trying to secure new lines of credit and to find a way to refinance Chrysler Financial loans that were soon to come due. The last thing the company needed was an attack by a corporate raider, which would scare away banks and kill any chance for new financing.

Within a few days, Kerkorian, accompanied by Alan "Ace" Greenberg, chairman of Bear Stearns, showed up at Iacocca's office in the Metropolitan Life building in New York.

"Lee, you're probably wondering why I invested over a quarter of a billion dollars in Chrysler," Kerkorian said. If he was angry or insulted about the poison-pill provisions, Kerkorian was disguising it well. "I invested in Chrysler because I believe in you."

Iacocca was pleased and flattered, though he was still unsure about whether Kerkorian posed a real threat. After Kerkorian and Greenberg left, Iacocca turned to Steve Miller: "Can you believe we were just sitting here with our biggest shareholder, and he didn't once ask us how business was?"

The uncertainty over Kerkorian's intentions added to Iacocca's headaches, not the least of which was increasing pressure from his directors to find a successor. With the departure of Greenwald, the imminent retirement of Bidwell, and the directors' lack of enthusiasm for Miller, only Bob Lutz looked like a reasonable internal candidate. But Iacocca was dead-set against Lutz, who, in his view, was ego-driven, a loose cannon with the press, and a lover of sexy cars who failed to understand that automakers lived by building cars for the average Joe. Besides, Bob was disloyal. Lutz had ridiculed and insulted him behind his back — so to hell with him.

Iacocca needed another candidate, someone to satisfy the directors and keep Lutz out of the picture. And if he didn't offer a name soon, he risked letting the board's nominating committee lose patience and put its own candidate forward. Grabbing a top contender from another Big Three automaker was a possibility. There were more than a few talented GM and Ford executives who had their sights set on the top but knew their chances were limited where they were and who might be willing to trade for a sure thing at Chrysler.

"Who else is out there?" Iacocca asked Bidwell, with whom Lee was close, even more so now that Bidwell's retirement in a few weeks would mean that he had no vested interest in who became chairman.

"What about Penske?" Bidwell asked. Now *that* was a name no one had tossed around before. Roger Penske. He of the regal silver mane, owner of a winning Indy 500 racing team, monarch of a $2 billion truck-leasing, car-retailing, and diesel-engine empire. Bidwell's mind raced: What about Roger Penske, chairman of Chrysler? Penske's talents as an entrepreneur were beyond question. He had started his career as a car dealer and built his businesses one at a time. He bought Hertz's truck leasing business and plastered his name next to Hertz's. He bought 80 percent of GM's diesel-engine manufacturing business and instantly turned a losing operation into a moneymaker, managing to make friends with the United Auto Workers in the process. True, Penske's Cadillac dealership in New York City had flopped, but he operated several successful auto retailing ventures in California, including the nation's biggest Toyota dealership.

Roger Penske was a dynamo, a tireless, one-man human-motivation seminar. He jetted from factory to office, from racetrack to dealership—checking, inspecting, meeting, calling on customers. His penchant for spit-and-polish cleanliness was legendary: before he got into auto racing as a young man, gleaming cars were thought to be a bit sissified. Moreover, his energy was contagious; everyone around him seemed to move a little quicker and stay a little more focused when he was around. Penske was the ultimate hands-on manager. Was there a way to accommodate that talent in an enterprise the size of Chrysler? Entrepreneurs often encountered difficulty in the staid, tradition-bound world of big corporations. Ross Perot had found General Motors to be uncomfortable with—even hostile toward—executives accustomed to making swift judgments and taking prompt action.

Iacocca instructed Bidwell to contact Penske discreetly, and a private dinner was arranged at the Kingsley Inn in Bloomfield Hills. The dinner went well. Penske was not just flattered but enthusiastic. Penske knew the auto industry and its executives intimately. He'd instantly divined that the meeting with Bidwell was to test his interest in Iacocca's job. Naturally, the idea of bringing an entrepreneur in to run Chrysler presented complications. Penske was the key figure in a privately owned empire that required his day-to-day presence. Chrysler would probably have to buy his company to get him, a transaction that might cost several hundred million dollars. If a buyout such as that wasn't a deal-killer, then Penske was *very* interested.

Quite by coincidence, Bidwell was scheduled to meet the next morning with members of the board's nominating committee to discuss succession. The three members of the committee, Malcolm T. Stamper, Jean de Grandpré, and Bob Lanigan, wanted to hear Bidwell's thoughts about Lutz, Miller, and other possible candidates. (Chrysler's vice president for personnel, Glenn White, who often

attended meetings of the nominating committee, was also present.)
Bidwell was bubbling with elation from the previous night's meeting
with Penske. To him, Roger Penske was the ideal solution to Chrysler's dilemma. If anyone in the world had the knowledge, the experience, and the instant credibility to run a big automaker it was he. To
top it off, Penske was an automotive marketer's dream. Bidwell and
Chrysler's advertising agency, Bozell, had used Iacocca to sell cars,
and they could do the same with Penske. The image of a handsome,
successful racing-team owner and a former driver was the perfect
antidote to Chrysler's less-than-stellar reputation with younger buyers. Indeed, Penske already had appeared in Rolex advertisements.
It wasn't hard to imagine him in front of the cameras, hawking Chryslers, Dodges, and Plymouths.

Bidwell started by offering the nominating committee his opinions
concerning Lutz's and Miller's suitability as candidates, their pluses
and minuses. Then he turned to his own brainstorm, Roger Penske.
As Bidwell expanded on the attributes he thought made Penske a
dynamite candidate, he could see Stamper, Grandpré, and Lanigan
warm to the idea. By the meeting's end Bidwell believed that the
directors were quite enthusiastic, and even eager to discuss Penske's
candidacy with Iacocca. Bidwell felt very gratified. He had worried
that Iacocca would not leave gracefully. The Penske solution had so
much to offer; perhaps Iacocca just might be receptive.

Throughout the spring of 1991 Iacocca and Penske met several
times. Bidwell, who felt pride of authorship for first having proposed
Penske, sensed the meetings were proceeding satisfactorily. Bidwell
was officially retired and working for Chrysler as a $100,000-a-year
consultant in charge of the daily-rental-car companies owned by
Chrysler. As a result, he maintained his office at headquarters and
kept in close touch with Iacocca. Bidwell told Iacocca he thought
Penske a spectacular choice for chairman and, by the same token,
that Lutz definitely was not. Earlier he had avoided revealing to the
chairman what Lutz had been saying behind his back. But then he
had been a Chrysler employee—and Lutz's boss—and felt some ob-
ligation to be discreet to avoid discord.

Iacocca was initially lukewarm about Penske. Bidwell was disappointed, but he realized he had committed a tactical error with regard to Iacocca's ego. Because he and the members of the nominating committee had been so enthusiastic, perhaps too enthusiastic, Iacocca now felt that he was losing control, that he was being pushed to like Penske. Feeling rushed, Iacocca's unwillingness to let go reasserted itself. He called Jerry Greenwald to tell him about Penske and ask him to do a little investigating for him: *Evaluate Roger Penske's business holdings and find out what they're worth.*

On May 27, 1991, Rick Mears drove a Penske-Chevy racer to victory in the Indianapolis 500, the eighth time that a car sponsored by Roger Penske had won the race. Penske was jubilant on the phone when Ben Bidwell called to congratulate him and find out how the talks with Iacocca were progressing. "Had to show the boss what we've got," Penske told Bidwell proudly, referring to Iacocca. Things looked promising. The two men discussed the possibility of bringing Mears and another Penske driver, Emerson Fittipaldi, to the annual meeting of Chrysler dealers in August—presuming that Penske was named Iacocca's successor by that time.

A few days later, Bidwell spoke to Penske by telephone.

"It's over," Penske said. He wasn't coming to Chrysler.

Bidwell was crushed. He couldn't imagine what had happened. Iacocca was evasive. He told Bidwell that Penske's corporations, based on Greenwald's valuation, were priced too high. Penske wanted $300 million; his business wasn't worth that much. In addition, Penske's diesel-engine manufacturing operations didn't fit in neatly since Chrysler didn't make the type of trucks powered by large diesel engines. Penske wanted to bring key executives with him into Chrysler management, including Bill Hoglund, an executive vice president of General Motors. He also wanted to fire a few people at Chrysler, including Bob Lutz.

The contemplated personnel changes didn't bother Iacocca. It was something else. Bidwell suspected that it was Penske who had done the backing-off. Iacocca then confessed he had wanted Penske to join Chrysler as chief operating officer, then become chief executive officer in 1992 with Iacocca staying on as chairman for another year. In other words, Iacocca wanted Penske to come aboard as number two and remain under his wing for eighteen months. Penske, who had started his business career as a Chevrolet dealer and built a fortune without having to answer to anyone, wasn't interested in playing understudy to Iacocca.

"Hey, when I came from Ford I agreed to play second fiddle to Riccardo for a while," Iacocca told Bidwell. "Penske can do the same."

The difference, Bidwell reminded Iacocca, was that in 1978 Iacocca didn't have a job. Roger Penske, as much as he might have wanted to run Chrysler, really didn't need the job. At that instant, Bidwell grasped what Jerry Greenwald had discovered in 1990—that it was unlikely that Lee Iacocca would be leaving Chrysler willingly.

(In Penske's version of events, accepting the top post as chairman of Chrysler was never more than a remote possibility. Penske, who was reluctant to describe the negotiations except to say that he had met with Iacocca three times, said he decided he couldn't consider the job seriously because of "prior commitments to executives and managers at Detroit Diesel.")

The American media in the fall of 1991 were filled with reminiscences of the fiftieth anniversary of the Japanese attack on Pearl Harbor. Consequently, it was a time of meditation on the nature of relations between Japan and the United States. The United States and its allies only a few months earlier had clashed with Iraq in the Persian Gulf War. The war, though short, had been an expensive undertaking that helped spark a recession, depressing the sales of cars and trucks and causing steep financial losses for U.S. automakers. Japan contributed financially to the war effort, but its support was slow in coming and included no personnel. Having foresworn militarism, Japan had little interest in sending troops halfway round the world. Some Japanese, recalling the horror of World War II bombing raids, sympathized with Iraq. Japan's ambivalence about Operation Desert Storm, and weak automotive sales angered some Americans and presented a rich opportunity for anti-Japanese demagoguery.

With Chrysler in a grim financial state, Iacocca seized the moment. The atmosphere was ripe to warn that "another kind of Pearl Harbor can happen again if Japan's aggressive economic policies continue." In truth, Iacocca was bitter and frustrated for many reasons that had nothing to do with Japanese trade. Chrysler appeared to be on the edge of an abyss, and the automaker's board of directors was pressing him to find a successor. "The limit of American tolerance is being reached," he wrote in his weekly newspaper column, sounding as though he yearned to press the button on the bomb-bay doors.

Recalling American martyrs of World War II, Iacocca attempted to rally emotions against Japan. "Most of the 5,000 Pearl Harbor survivors in Honolulu for the anniversary will be staying in hotels owned by Japanese," he wrote. "They won't have a choice. And the irony won't be lost on them as they toss their wreaths on the water over the battleship Arizona." Iacocca wasn't alone. Political posturing in Washington and Tokyo was peaking a year before the U.S. presidential elections. Jobless and trade-deficit numbers were potent weapons in the hands of a Democratic challenger to George Bush. Every presidential candidate had at least to sound tough on Japan, which is what soft-spoken Paul Tsongas must have been thinking as he cracked, "The Cold War is over, and Japan won."

Iacocca had been reminded once before that his rhetorical flourishes against Japan were insulting, and frequently injurious, to the nation's 7.3 million Asian-Americans. In a 1985 speech to the House Democratic caucus in White Sulphur Springs, West Virginia, Iacocca had suggested telling the Japanese prime minister that he must take steps to reduce our trade deficit or the U.S. Congress would take action. Punctuating his point, Iacocca used the word *sayonara*. He didn't realize that *sayonara* used in that context would strike a Japanese listener as offensive. Robert T. Matsui, a third-generation Japanese-American and a congressman from California, was present at Iacocca's speech (which had, in fact, been written by Ben Bidwell).

"My feeling was, when a prominent person like Lee Iacocca gives a speech like that—he's credible—it could affect my son, my family and those like us," Matsui said later. "I can't sit back and tolerate that." Denying that he had said anything derogatory, Iacocca declined to withdraw his remarks but sent Representative Matsui a conciliatory note.

Asian-Americans, including a few Chrysler employees, had written to Iacocca intermittently to complain of his anti-Japanese rhetoric, but he steadfastly denied being racist or xenophobic, saying his only quarrel was with U.S. policy toward Japan. In the wake of rising violence against Asian-Americans, the U.S. Civil Rights Commission report concluded that "Americans take out their frustrations about Japan's economic success" on all Asian-Americans without regard to their national origin. The civil rights agency called on presidential candidates to avoid remarks that could be construed as "race-baiting" or "Japan bashing."

Reacting to speeches directed against Japan, Malcolm Wallop, Republican senator from Wyoming, denounced what he saw as cockeyed economic thinking and sinister portayals of Japan. "America, not Japan, is the world's largest exporter," he wrote in an essay published in the *Christian Science Monitor* in 1992. "Why do we hear so much about the Japanese 'buying up America' when Japanese investment is still only 60 percent of total British investment in the U.S.?" Reminding readers that Japanese companies in the United States employed 600,000 workers, Wallop wrote, "it's time to stop looking at Japan as the enemy and start recognizing it as one of America's vital trading partners."

Those who regarded Japanese businessmen as the advance guard of a mighty conspiracy to take over and control the U.S. economy could read their beliefs between the lines of a best-selling work of popular fiction. In Michael Crichton's *Rising Sun*, a 1992 police thriller set in Los Angeles, Japanese businessmen tried to steal American technology, committing murder and suborning a U.S. senator in the process. As the police closed in on the fictional Nakamoto Corporation, its employees insisted they were victims of racism. Not even the press could be trusted because Nakamoto had been busy buying off reporters.

In *Rising Sun* Japanese businessmen were caricatured as a diabolical horde of faceless automatons, striving to bring America's economy under Japanese control and to use their national wealth to buy up American real estate and other assets. Americans, by contrast, were portrayed as weak, naïve, undisciplined people working in industries that just seemed to be losing their vigor. Crichton's renown as a novelist assured his book a wide audience, even though a number of reviews criticized its faulty economic analysis, racial stereotyping, and numerous factual distortions. Had Crichton's errors been incidental to producing a police thriller, they might not have been so disturbing. But Crichton's book, by his own admission, was written to warn the nation, not entertain it. In an afterword to the novel, Crichton cautioned that it was time "for the United States to wake up, see Japan clearly, and to act realistically."

"The Japanese invited thousands of experts to visit—and then sent them home again," wrote the Harvard-educated author. "We would do well to take the same approach. The Japanese are not our saviors. They are our competitors. We should not forget it." Could an intelligent, articulate author truly be suggesting that the government expel the Japanese, as the Spanish had done with the Jews during the Inquisition?

Crichton's not so veiled suggestion that the Japanese should be thrown out received a sympathetic hearing in many quarters. As economic conditions grew difficult in the recession following the Gulf War, Asian-American and Japanese-American cultural organizations throughout the country reported an alarming increase in the number of racially motivated incidents. Some were minor: a bottle thrown from a car; a racial slur at a shopping mall. Others were personal assaults and crimes against property. A spokesman for the San Francisco–based Japanese American Citizens League said his group feared that someone was going to be killed. No one had forgotten Vincent Chin.

In February 1992, a Japanese businessman was found murdered in his home in Camarillo, California. Yasuo Kato, forty-nine, was stabbed twice in the chest with a large hunting knife while unloading groceries in his garage. Ventura County sheriff's deputies had one important clue: just a few days before he was killed, Mr. Kato called the sheriff to report that two men on a motorcycle had come to his

home, knocked on the door, and demanded money. The men told Kato
they were entitled to money because the Japanese were responsible
for America's economic slump.

A former collegiate martial arts champion, Kato shoved the men
out of his house, the report said, which made them angry. The deputies had promised to patrol in the neighborhood.

Two years passed, and the murder was still not solved. Asian-
American groups and the Japanese embassy were cautious about con-
necting Kato's death to anti-Asian sentiment, but his confrontation
with the two motorcyclists was reason enough to put all Asian-Americans on guard.

If anyone should have been ultrasensitive to the plight of immigrant Americans who yearn for acceptance and an end to discrimination it was the nation's most prominent businessman, a man who had
grown up amid slurs against "wops" and "dagos" in Allentown, Pennsylvania. One of Lee Iacocca's favorite stories was about the time
anti-Italian schoolmates in the sixth grade had stuffed the ballot box,
stealing from him the election of captain of the safety patrol. "Being
exposed to bigotry as a kid left its mark," Iacocca wrote in his autobiography. "I remember it clearly, and it still leaves a bad taste in my
mouth."

ENDGAME

A HALF-HOUR'S DRIVE due north of Detroit on Interstate 75 lies the rural community of Auburn Hills, Michigan. Cruising through the shady, bucolic glens of Auburn Hills, the mind drifts far from the auto industry's gritty industrial scene. Here, in 1991, far removed from the urban decay surrounding Chrysler headquarters in Highland Park—the tiny city enclosed within the city of Detroit—a new $1 billion technology center neared completion. In time, the complex would house Chrysler's new corporate headquarters, replacing the automaker's longtime home in grim, crime-infested Highland Park.

The sparkling ultramodern technology center was a monument to Lee Iacocca's tenacity and to the totality of his rule. Every other executive at Chrysler, worried by the distractions and financial obligations of a big construction project, favored abandoning the Chrysler Technology Center or at least radically curtailing its size. It became known among them as "Iacocca's Mausoleum." At a seminar in West Palm Beach in April 1989, held to introduce the latest in "New Age" communication and decision-making skills, Chrysler's two dozen or so officers agreed that the construction project was too expensive. Chrysler was, after all, trying to reduce costs. The annual cost-cutting target had grown, at Iacocca's insistence, from $1 billion to $1.5 billion and now to $3 billion.

When Iacocca arrived for the last day of meetings in West Palm, Doug Anderson, who was leading the conference, mentioned that the officers were worried about the cost of the tech center. They thought it should be cut, some thought completely, and others from a $1.5 billion to a $1 billion project.

To everyone's surprise, Iacocca agreed. The tech center was the future, he said, and it had to be completed. But it could be smaller; just so long as it was built. As he had during the discussions on whether to buy American Motors, the chairman wielded his clout to move unilaterally on what he saw as a fundamental issue. Though he paid lip service to the importance of listening and discussing, Iacocca had no time for "consensus management." Discussion was necessary, but at times he had to take decisive action. In this case, Iacocca insisted that Chrysler needed to close its numerous small installations scattered throughout Detroit as soon as possible and to concentrate everyone in a large, safe edifice. Engineers wasted untold time and money traveling back and forth between offices. Besides, recruiting young talent from the East or West coasts to Highland Park was difficult, if not impossible, if they couldn't offer reasonably pleasant surroundings.

Less than a mile from the magnificent structure being erected in Auburn Hills, in a low, nondescript building, the 744-member L/H team — shrunken slightly from its original 850 members — was racing toward its deadline. The lights in the team's temporary building burned far into the night; many of its cubicles were filled on Saturdays and Sundays. The team really was, in fact, an agglomeration of many smaller sub-teams. Discussions among these groups were long and frequent, planning was meticulous. Consequently, decisions seemed to take longer than usual, but when action was taken, it unfolded very quickly and rarely had to be corrected. The Japanese called the process *nemawashi*, which meant "stimulating the small roots to grow." Power trips and ambition were discouraged, as was meddling from Chrysler executives. When "suggestions" arrived from headquarters, Bob Lutz instructed Glenn Gardner to ignore them — politely. Lutz, Castaing, and Gardner were meeting often, watching the day-by-day progress of the L/H, which, without a doubt, had evolved into the project on which Chrysler's entire future was riding.

Reliance on consensus management had succeeded in slicing

many months of development time from the typical fifty-four-month schedule. With less than a year to go until the summer of 1992, the L/H was coming together quickly, perhaps in as few as its planned forty-two months. For the first time in its history, Chrysler engineers were thinking extensively about how to manufacture the car they were designing. With workers from the factory serving on the team, parts were designed so that they could be assembled safely, quickly, less expensively, and at a high level of quality. Several reviewers from car magazines were invited to Chrysler's test track in Chelsea, Michigan, to drive early prototypes of the L/H sedans and to evaluate engine and handling performance. Automakers rarely allowed outsiders to see their creations in advance—unless, that is, the cars they were currently selling were so awful as to threaten the maker's survival. First reactions to the L/H were encouraging. To believe that the L/H was destined for widespread consumer acceptance and popularity was Glenn Gardner's only option. The alternative, L/H's failure, suggested a future too grim to contemplate.

Sadly, the faith of lenders, upon whom Chrysler's existence depended, was rooted in the here and now. In mid-1991, Chrysler had about $2.8 billion in the bank. For most companies this would be a tremendous sum, but not for an automaker. Chrysler needed a minimum of about $800 million in the bank for daily operations, and it expected a $1 billion net cash outflow for 1991, due to weak sales. With a cushion of only $1 billion—an amount that might be depleted in less than six months if bigger rebates became necessary—Iacocca needed to secure a $1.75 billion emergency line of credit from a consortium of banks. Bankers, although they love to collect fees from emergency lines of credit, don't like the idea of actually *lending* money for an emergency. They had questioned Chrysler's finance executives closely: Were they sure the L/H was going to be popular? Who was going to run Chrysler when Iacocca left? Jerry York, who had succeeded Steve Miller as chief financial officer when Miller became vice chairman, insisted that the new Jeep and L/H looked terrific and that several seasoned executives, including Lutz and Miller, stood at the ready to step in for Iacocca. The bankers granted Chrysler the emergency credit, praying it would never be needed.

Chrysler's other pressing worry was Chrysler Financial. Although the subsidiary was financially much more secure than the parent corporation, its ratings reflected the shakiness of the parent. Credit analysts could only assume that in a crunch, the corporate parent and the subsidiary would help one another and were, in fact, one. Thus, the weaker of the two ratings—in this case, that of the parent—became the rating for both. Chrysler Financial needed to refinance several billion dollars of obligations and suddenly, because of low credit ratings, it was shut out of the commercial paper market, where it had previously raised money by selling securities.

Credit analysts told Chrysler Financial officials that Chrysler might improve its ratings by raising more capital. Selling 56 million shares of stock ten or eleven dollars a share was not something Lee Iacocca hankered to do—especially since Chrysler had bought back $1.8 billion worth of stock just a few years earlier at prices above twenty dollars a share. Selling shares diluted the stock, depressing the price. Chrysler, however, had little choice. Iacocca, Lutz, and Miller, along with the investment bankers, embarked on a "road show" to potential investors in Europe and the United States to reassure them that buying new Chrysler stock was a safe bet. Privately they explained that Chrysler was fine. The company was selling stock only to satisfy credit analysts. The road show was a carefully scripted display of solidarity, a device to project confidence in Chrysler's future.

In reality, the atmosphere inside Chrysler had never been filled with more dissension and distrust between Lee Iacocca and Bob Lutz—or between Iacocca and Chrysler's directors.

Iacocca's reluctance to consider Lutz as Chrysler's next chief executive solidified into an iron wall of resentment. Lutz, in part, had brought it on himself. Inevitably, the remarks he was making about his boss got distorted and embellished as they were repeated. Confidants of Iacocca told him that Lutz had called him "the guinea who runs this place." Though Lutz denied the ethnic slur, the story had already entered the realm of corporate folklore. By letting everyone know how little regard he had for Iacocca—and doing it so indiscreetly—Lutz had antagonized Iacocca in a manner that was extremely difficult to patch over.

Iacocca never confronted Lutz directly about the reported slurs. Instead he singled out Lutz's principal corporate ally, François Castaing, for hostile treatment. Iacocca held Castaing responsible for the mail he had been receiving from Chrysler engineers urging him to restore the old organizational structure. Several of the letters charged that millions of dollars were being wasted and that the L/H and other future models were falling behind schedule.

After reading the letters, Iacocca summoned Lutz and told him to fire or at least demote Castaing. Lutz refused.

"Tell me what you have against him," Lutz said to Iacocca.

"I'm hearing things about him," said Iacocca, "and I want him out of here."

"If you want me to fire someone, I've got to have specifics," Lutz said.

"I don't need specifics," Iacocca roared.

Lutz warned Castaing that Iacocca was on the warpath, though there was little he could do except encourage Gardner to finish the L/H with all due speed. At times Lutz wondered if Iacocca wasn't preparing to fire both him and Castaing. Early in 1991, Lutz and Iacocca met for a dinner at a little restaurant near Detroit to try and clear the air. Instead, Iacocca exploded, loudly accusing Lutz of being not nearly as clever or as attuned to the tastes of car buyers as he fancied himself to be. Lutz ought not laugh too loud at the Imperial and New Yorker, Iacocca said, because plenty of buyers were still devoted to those cars. By the next day, the blowup was being whispered about up and down the corridors at Chrysler headquarters.

Iacocca's antipathy for Lutz centered more on style than specific instances of disloyalty. Lutz's father had been a banking big shot on the board of Crédit Suisse, and Iacocca's father a smalltown hotdog vendor. Lutz possessed flair and self-assuredness while Iacocca was haunted by insecurities. Lutz spoke French and German and sprinkled his conversation with quotes in Latin; Iacocca was able to mumble a few Italian phrases that he had learned as a child. Iacocca grew annoyed when people spoke a foreign language in his presence. Once or twice when Castaing and Lutz had spoken French within Iacocca's hearing, he'd snapped, "Cut out that French shit!"

Extremely galling to Iacocca was the fact that Lutz had constructed a network of supporters throughout Chrysler, on Wall Street, and even among the directors. So far, Lutz had proven himself by reorganizing engineering, building the Viper, and redesigning the L/H. His charisma and leadership qualities were impossible to deny. A strong case could be made that Lutz had earned the chairman's job, that he, not Iacocca, deserved the lion's share of credit for the next wave of new vehicles. Iacocca might have saved Chrysler in the 1980s, but his neglect had caused Chrysler to stumble again into crisis. Lutz was now the one preserving Chrysler's chance for a future.

If Lutz became chairman, Iacocca feared he would tell everyone that Iacocca had nearly ruined Chrysler and had to be pushed out. Lee had reasons to stick around—at the very least to stop Lutz from spoiling his place in automotive history. And Iacocca knew how easily that could be achieved: his own autobiography had gone a long way to enlarge the distorted portrait of a drunken, womanizing Henry Ford who had booted him out of Ford for no reason.

Iacocca was convinced that he must hang on to his office at least until the L/Hs and Jeep Grand Cherokee were introduced in 1992. If the vehicles were hits, he would take the credit. To secure his position, Iacocca had to prevent Lutz from being named chairman. Therefore, he had to show the directors that Lutz was a bad choice. Iacocca's opinion still carried tremendous weight, but it might not be enough in light of Lutz's considerable achievements and his following within the company. Politically savvy as he was, Iacocca knew that he must categorically disqualify Lutz. Iacocca called Red Poling, Ford's chairman, to ask what he thought about Lutz as a potential chief executive officer. Don Petersen, Poling's predecessor, was also approached for his opinion.

Of course, Iacocca knew in advance that Poling and Petersen were at best ambivalent about Lutz. Iacocca had exploited Lutz's precarious standing with the two Ford executives in 1985 to help convince him to leave Ford and join Chrysler. Iacocca knew that Lutz had fallen into Poling's bad graces and didn't wish to work for him any longer. Poling had criticized Lutz for concentrating on cars he liked in favor of cars that made money. Petersen, Ford's chairman in 1985, had been a supporter and admirer of Lutz but ultimately sided with Poling and other Ford executives who believed Lutz's love of racy cars and his career ambitions sometimes clouded his judgment.

Neither Petersen nor Poling was willing to give Iacocca a written evaluation of Lutz. The Ford executives were wary of being drawn into a succession struggle at a competing automaker. In any case, executives at Ford who heard that Iacocca was collecting information on Bob Lutz thought it highly odd that Chrysler's chairman had come to Ford for an evaluation. Preserving his grip on power while neutralizing Bob Lutz became Iacocca's obsession. If Iacocca had wanted to step aside in early 1991, he would have had some attractive options open to him. In May 1991 the governor of Pennsylvania, Robert P. Casey, offered him the senate seat of the late John Heinz. The Democrats wanted a strong figure who was sure to be reelected; and Iacocca had always professed interest in public service. But he turned down the offer, citing his age, family, and "corporate obligations." The obligations undoubtedly included finding a successor other than Lutz. In fact, he called the effort "ABL"—for "anybody but Lutz." Bit by bit, he gathered testimony and anecdotes with which to convince the board's nominating committee that Lutz was the wrong choice for chairman. There were others who had known Bob Lutz and had reservations about him. Mike Hammes, Chrysler's former international vice president, who had worked for Lutz at both Ford and Chrysler, was another who thought Lutz put his own career ambitions before the interests of the company. Hammes was an Iacocca loyalist and, though he no longer worked at Chrysler, was prepared to testify if Iacocca needed him.

Within a few weeks of Roger Penske's decision to drop out of the running, the directors stepped up their pressure on Iacocca to propose a suitable successor. In late August of 1991 the *Detroit News* published a story in which Iacocca suggested the opposite, that there was no big rush. He was quoted as saying that he was only sixty-four years of age, not sixty-six, and that the directors weren't going to begin looking for a successor until 1992. The story completely blindsided the outside directors, and several reportedly said so at Chrysler's early September board meeting. As far as the board was concerned, Iacocca had delayed naming a successor for too long. More than two years had passed since Jerry Greenwald's departure. The directors had been disappointed by the collapse of talks with Penske a few months earlier, but they decided that the reasons had little to do with Iacocca's reluctance to retire. Finally, enough was enough: at the meeting, the outside directors insisted that Iacocca clear the air.

On September 5, 1991, Chrysler issued an announcement stating that Lee Iacocca had agreed to retire from the company at the end of 1992. Despite the directors' desire to avoid open hostilities with him, a succession plan was crucial. With Chrysler about to sell fifty million shares of stock, and with credit analysts uneasy, the directors wished to avoid a management debacle caused by a chief executive who was conflicted between his desire for immortality and the responsibility to provide leadership continuity.

According to the statement, Iacocca agreed to stay on as chairman and "help the board" find a successor. In case the public had been wondering about Iacocca's quotes in the newspaper the previous week, the statement also said "the board has already begun discussions on the selection of a successor." The announcement was meant as a signal to the outside world that Chrysler directors had seized control of succession, which until that moment had been Iacocca's domain.

Until a new chief executive was found, Lutz's candidacy remained alive, at least theoretically. Iacocca, on behalf of the board, had invited Steve Miller and Bob Lutz to write letters to the board's nominating committee back in the summer of 1991, in effect to describe themselves and their qualifications as candidates for chief executive. The board's nominating committee consisted of Stamper, Grandpré, and Lanigan, but Iacocca had been largely directing the search himself.

Steve Miller, though technically a candidate for chairman, was a long shot. Although Miller and Lutz had each stated publicly that they were prepared to serve as each other's subordinate, few if any directors considered Miller a leading contender. Bright and energetic,

Miller tended to be quiet and reserved, lacking the charisma or flair Iacocca thought necessary to lead an automaker. Miller also had little experience with Chrysler's basic products, cars and trucks. He had grown up on the finance side of the business and had never gained the general management, marketing, or operating experience that might have prepared him for a chief executive's job. Miller's wife, Maggie, didn't help his cause much, either. As a corporate spouse she continued to stick out, regularly blurting out her thoughts and opinions with little regard for the time or place. Now and then she would jam a stogie in her mouth and light up. Executives and directors—though some liked Maggie personally—braced themselves when she approached.

In his letter to Iacocca, Miller expressed his desire to be named chief executive. Conceding that he didn't have "gasoline in my veins," he observed that a gentler leadership style might be conducive to a stronger team spirit. Miller proposed that he be named chairman and chief executive, with Lutz as president and chief operating officer and Jerry York as chief financial officer. Miller urged the board, whatever it did, to select an internal candidate rather than someone from the outside. Furthermore, he suggested to Iacocca that he forgo a prominent role on Chrysler's board after his retirement, since his overpowering presence might prove cumbersome for his successor.

As Iacocca had requested, Miller enumerated the reasons why he thought Chrysler was floundering. The stock repurchase plan had been a mistake. Selling a large number of cars to daily-rental fleets had been self-defeating and unprofitable. Investment in light trucks, though substantial, had been insufficient. The "romance of foreign partnerships," shrouded in fancy accounting, had blinded Chrysler to financial losses. Chief among his recommendations was that Chrysler return to its roots in the automobile business. Miller also urged that Chrysler stop bashing Japan, and that it get rid of weak brands like Plymouth and repair damaged relations with the union.

By the time Lutz wrote his letter, Iacocca's unrelenting hostility toward him and François Castaing made him a very dark horse. In his letter, Lutz described his European roots and his multifaceted career at GM, BMW, and Ford. Lutz also disclosed in the letter something that had become an open secret in Detroit: his marriage to Heide had fallen apart. (Though Lutz did not elaborate in the letter, his estrangement and the divorce proceedings had turned ugly. The Lutzes, through their lawyers, were fighting over who owned the $1 million estate near Ann Arbor and who should live in the house during the divorce proceedings. Heide Lutz's attorneys, attempting to exert leverage, had served subpoenas on Chrysler and on Lutz's secretaries, demanding to see travel records of trips with anyone other than clients and expenses reflecting purchases of flowers, jewelry, and other gifts.)

Lutz wrote that "a single-minded dedication to outstanding cars and trucks" was the most important ingredient for the success of an automaker. Rather than regarding quality as freedom from defects, he stressed that quality must be understood in a new light, as "things gone right," features and characteristics that make the driver feel secure and in control. Paying attention to the growing market that buys Accord and Camry versus the declining market of traditional models and "having the courage to let our platform teams execute is the number-one element in our rebirth."

Lutz's letter was palpably devoid of optimism or spirit, a reflection surely of his belief that unless Lee Iacocca suddenly dropped dead, his chance of being named chairman was virtually nil.

From the moment Iacocca quietly hatched his plan to bring Jerry Greenwald back to Chrysler as chairman, Lutz, Miller, and the few others who heard of it were upset.

Greenwald, ever the diplomat, had carried out Iacocca's orders faithfully for more than a decade, never contradicted him in meetings, and always spoke respectfully of his boss, particularly if someone who might report back to Lee was in earshot. In Greenwald's view, there was no other way to handle an executive of Iacocca's authoritarian bent, one who also enjoyed 110 percent support from his board of directors and unquestioning public adulation. To many, Jerry Greenwald appeared to be Iacocca's number-one yes-man; he himself knew that he was regarded in that way. Greenwald often confronted Lee privately and in such situations wasn't afraid to speak his mind, but he knew that Iacocca's ego simply could not bear direct confrontation in front of others. Greenwald knew that the only way to change Iacocca's mind was to speak to him, sometimes sharply, but only when others weren't around. Now and then Greenwald had no choice but to carry out orders he disagreed with. In those cases, he made clear that they were the boss's orders, not his.

As Jerry Greenwald and Lee Iacocca discussed the possibility of Greenwald's return through the fall and early winter of 1991, Lutz, Miller, and some of the directors wondered if Greenwald wasn't being positioned to play the role of Charlie McCarthy to Iacocca's Edgar Bergen. Bringing in Greenwald gave Iacocca a chief executive whom he could influence and manipulate, and thereby preserve his own power.

Greenwald was eager to return to Chrysler. The failure of the United Airlines deal had added $9 million to his bank account but left him without a high-profile position. Making deals on behalf of Dillon Read for little-known companies was tough going and too anonymous. Greenwald enjoyed the excitement of operating in the limelight, under the scrutiny of the financial community and the press. The Chrysler bailout had given him a taste for "making history," something he wanted to achieve again. Iacocca forgave him for leaving; all that remained was the approval of Chrysler's board.

The Chrysler directors, however, were divided over the question. A few of them wondered, as did Lutz and Miller, if Greenwald was not merely a stalking horse for Iacocca's ambitions. Lee was talking about his "continuing role" at the automaker after his retirement. Directors weren't sure what that meant. As a nonexecutive chairman of the board, or as chairman of the board's executive committee, Iacocca might effectively overshadow Greenwald or, for that matter, any other chief executive. Moreover, Greenwald's return might provoke bitterness in those who had stayed and persevered after Greenwald jumped ship.

Greenwald began to call directors to request their support for his candidacy. Aware that he might be perceived as Lee's proxy, he emphasized his desire for no more than a short, friendly overlap with Iacocca prior to Iacocca's retirement. Iacocca had no justification to propose a long "break-in" period, as he had with Penske, Greenwald said. Greenwald had been Chrysler's number-two executive and knew the place well.

However, in the fall of 1991, the Chrysler board was a much different, more restive body than the one Jerry Greenwald had known. Directors had grown impatient with Iacocca's failure to come up with a succession plan. The board suffered a major embarrassment in the spring of 1991 when details of how Iacocca had squeezed from the company an additional 250,000 shares of Chrysler stock and other compensation — including payment for two of his homes — in return for agreeing to stay as chief executive became public knowledge with the disclosure of the proxy statement. The board was becoming split into those within Iacocca's inner circle and those outside it. Some directors — Earl G. Graves, publisher of *Black Enterprise* magazine, Kent Kresa, chairman of Northrop, and Peter A. Magowan, chairman of Safeway — were upset by Iacocca's tendency to confide in his small kitchen cabinet of Grandpré, Lanigan, and Sticht. Directors who were excluded understood that Iacocca had been relying on it to exclude participation by Owen Bieber, president of the UAW — but once Bieber was gone, there seemed no justification for Iacocca to speak with only a few directors.

A serious obstacle to Iacocca's plan to install Greenwald as chief executive was the active opposition of Bob Lutz and Steve Miller. Lutz and Greenwald had not been on good terms at the time of Greenwald's resignation. Lutz bridled at taking orders from a "bean-counter," an executive who understood cars from an analytical, "left brain" perspective. Greenwald, as Lutz's boss, hadn't appreciated the challenges to his authority. Lutz, never one to hide his thoughts, put the word out to directors that "if Jerry comes back, I'm history." Miller too was opposed to Greenwald (whom he personally liked and respected) because he believed Chrysler couldn't afford to lose Lutz.

Thus, Iacocca tried to persuade directors to accept Greenwald in the fall of 1991 while remaining dead-set against Lutz. Bob, though he wanted the job, was willing to work for Miller, but not for Greenwald. Miller wanted the job and would work for Lutz or Greenwald, but believed bringing Greenwald back was a bad idea. The directors, the individuals who were legally responsible for management succession, were divided among themselves and somewhat wary of Iacocca's intention to stay on as a force at Chrysler. Several were convinced that a strong outsider was the only logical solution.

The search had deteriorated into a stalemate, snarled by a morass of ambition and recrimination. The company had little time to lose, for its financial condition was deteriorating by the day. If the directors didn't act soon to untangle the mess, Chrysler might never have the chance to sell the Jeep Grand Cherokee and the L/H cars that were nearing the start of production. Early in September, Glenn White, Chrysler's personnel chief, informed Steve Miller that a group of outside directors wished to meet with him on September 30 in New York to discuss management succession. Miller wasn't sure how many directors were going to attend the meeting. He was heartened, though, that the board finally appeared to be seizing the initiative and acting independently.

The meeting was to be a rare and welcome opportunity for Steve Miller to say what was in his heart without Iacocca listening. He decided to summarize his thoughts on paper. With uncharacteristic frankness, he wrote that to date the search for a successor had been a "sorry chapter" in which Iacocca dominated a process that should be the responsibility of outside directors. The ambivalence and delay had upset lenders and demoralized company managers, who felt the competence of executives was being called into question.

Then Miller unloaded on Iacocca. With a nod to his accomplishments, Miller argued that Iacocca under no circumstances ought to be invested with a "continuing role" at Chrysler, since to do so would compromise his successor's ability to chart a new course. Iacocca, his talents notwithstanding, had made several key blunders. They included the share buyback, the investment in Maserati, an ill-fated joint venture with Renault, and Chrysler's involvement in daily-rental fleets. He criticized his boss for issuing contradictory orders and intimidating subordinates.

Without mentioning names, Miller wrote that a top Chrysler dealer, claiming to speak for many others, had demanded Iacocca's retirement. One of Chrysler's ten biggest lenders had sent an executive to Miller to say there would be no new financing unless Iacocca retired. Mitsubishi wanted Iacocca to retire. The hatred between Iacocca and Owen Bieber was damaging labor relations. Washington was fed up with Lee's tirades. Many employees were discouraged by his insistence on leading a "regal lifestyle" during a period of austerity. Miller singled out the $2 million renovation of Chrysler's suite at the Waldorf as an egregious example of extravagance. Indeed, Miller noted that dozens of Chrysler employees had been kept on the payroll for no other reason than to tend to Iacocca's personal needs. He suggested the board pay Iacocca whatever he wanted within reason, regardless of the potential flak from the press and from shareholders, in order to effect "a complete separation." Iacocca's well-deserved reputation should be preserved and burnished.

Miller repeated his belief that he was the best candidate for the job. As for Greenwald, he had always been so closely identified with Iacocca that, in Miller's view, "none of us were sure what Greenwald really stood for." After himself, Miller thought Lutz to be the ablest candidate. Moreover, he wrote that Lutz's abilities as a "car guy" were vital for any new chief executive. But Miller wrote that he was distressed by what he termed a "witch hunt" and a "whispering campaign" carried out by Iacocca to impugn Bob Lutz's integrity. There was no evidence to support any attack on Lutz's character, Miller said; in fact, the opposite was true. Lutz's background and multilingual talents conveyed a positive image.

Miller sought to allay concern that, as chief executive, Lutz intended to purge executives who had been loyal to Iacocca and surround himself with cronies. Only one area of the company stood in jeopardy of a housecleaning. Lutz had preached dissatisfaction with Chrysler's sales and marketing capabilities. Both he and Miller believed that Chrysler's advertising relied too heavily on cash rebates and on corny hucksterism. Both men had opposed the promotion of Ted Cunningham to executive vice president for marketing, Ben Bidwell's old job. Iacocca had overruled them. Permitting Iacocca's influence over marketing to continue in the person of Ted Cunningham would be a mistake, Miller wrote.

He went on to criticize what he viewed as Iacocca's unfounded hostility to François Castaing and Lee's footdragging in firing Dick

Dauch and replacing him with Dennis Pawley. Lutz was well suited to lead Chrysler, Miller said. "My conclusion about Lutz is that there is nothing wrong with him that Lee's retirement wouldn't fix." He suggested that he was amenable to Lutz being named president and chief operating officer under his own chairmanship.

Iacocca won't like this solution, the letter said, but the board must decide whether to impose its own choice at the risk of a confrontation or, alternatively, to permit Iacocca to call the shots for years to come through a self-appointed, loyal successor such as he envisioned Jerry Greenwald being.

A few months earlier, Steve Miller might not have expressed himself so freely. He knew that the penalty for crossing Iacocca could be instant termination. By the time he wrote his letter, however, Miller believed the outside directors were asserting their primacy over executive succession and, indeed, over the chief executive's authority. Iacocca's retirement announcement earlier in the month had been achieved at their insistence. True, Lee was not gone and might still figure a way to stay behind the wheel at Chrysler, but a promise to retire at the end of 1992 was something. Thus, when Miller arrived at his meeting with the outside directors in New York on September 30, he presumed that what he had written was only for the eyes of nonemployee directors. Present at the meeting were five of the ten outside directors: Malcolm Stamper, Joe Antonini (the chairman of Kmart), Bob Lanigan, Jean de Grandpré, and Joe Califano (the former secretary of Health, Education, and Welfare).

Miller discussed with them his beliefs about succession, Iacocca, Chrysler's problems, and his own qualifications. He then handed them copies of the letter, which elaborated his views in detail.

Within the week, Miller received a call from Iacocca, summoning him to his office.

"How was the meeting?" Iacocca asked. (So, he had heard about it.)

"Fine, I told them what I thought," Miller said.

"Did you give them anything in writing?" Iacocca asked. Miller replied that he had.

"I want to see it," Iacocca demanded.

Miller provided Iacocca with a copy of the letter.

Iacocca was furious. He felt betrayed and undercut. At a moment when his relations with directors were at a low point, the last thing he needed was first-person testimony from a close subordinate that he was over the hill, abusing the company expense account, and in need of replacement — and recommending Bob Lutz as the best candidate!

Iacocca demanded that Miller destroy all copies of what he referred to as "the black letter," including Miller's own. He didn't want anyone gossiping about dissension at the top, and he certainly didn't want his historical legacy sullied.

"Are you going to destroy yours?" Miller asked. No, Iacocca replied. "Then neither will I," Miller said.

Except in the unlikely event that the board decided to name Miller chief executive immediately, Miller realized his career at Chrysler had ended the moment Lee Iacocca discovered the existence of "the black letter."

The public offering of thirty-three million shares of Chrysler stock in October 1991 was a crucial test of confidence for Salomon Brothers, as well as for Chrysler. The investment banking concern had, only weeks before, been rocked by a scandal that engulfed the firm's government securities trading department. The scandal ultimately cost John Gutfreund, Salomon's chairman, his job. Chrysler elected to stick by Salomon, the lead underwriter for its stock, rather than switch to another investment bank.

In the ballroom of the Waldorf-Astoria Hotel, the last stop on Chrysler's worldwide road show for potential investors, Iacocca railed about the government's mismanagement of economic problems, calling for a stronger stand against Japan. It had become a tired and familiar song, and everyone knew the lyrics. Portfolio analysts and mutual-fund managers listened to Iacocca flog Chrysler's soon-to-be introduced L/H, the Grand Cherokee, and the Viper, whose reception by consumers would be a determinant of the return on their investment.

Happily for Chrysler, demand for the stock was strong. On October 1, the entire offering sold out, albeit at a price of only slightly more than ten dollars a share. The additional $350 million generated by the sale of stock was far less important than the display of confidence that the credit analysts wanted to see. These experts, who evaluate the likelihood that a company is able to repay its debts, were heartened that investors were willing to risk fresh capital.

Meanwhile, workers at the brand-new Jefferson North Assembly plant in Detroit were receiving last-minute training in the assembly of the Jeep Grand Cherokee, the new sport-utility vehicle scheduled to go on sale in April. Before the start of production, the new machines and their computer controls had to be debugged. Shortly after the start of production at Jefferson North, assembly of the L/H sedans was scheduled to begin at Chrysler's assembly plant in Bramalea, Ontario. Members of Glenn Gardner's L/H team were shuttling feverishly back and forth between Auburn Hills and Bramalea, checking tools, programming computers, and inspecting prototypes to make sure the fit and finish of the L/H was equal to the promise of its innovative design.

The two Chrysler assembly plants were among the first in the U.S. auto industry to employ extensively the principles and techniques that Jim Womack and the MIT researchers had described in *The Machine That Changed the World.* Chrysler engineers, including Glenn Gardner, shepherded their transfer from Mitsubishi's factories in Japan and the Diamond-Star joint venture factory in Bloomington, Illinois. The Chrysler plants had just-in-time inventory systems, *andon* cords for stopping production to correct quality problems, and scores of other concepts and tricks from Chrysler's Japanese partner that were becoming routine features of the new Chrysler factories.

At the same time, Chrysler was taking steps to loosen its financial ties with its Japanese partner. Starting in 1989 Chrysler had begun to unload its remaining stake in Mitsubishi, which dated to the beginning of the partnership in 1970. Mitsubishi executives almost certainly would have been insulted had they not recognized that Chrysler desperately needed the cash, which totaled a tidy $310 million. The second step in unwinding the formal relationship was the sale of Chrysler's stake in Diamond-Star, which had been its learning laboratory. Because the Diamond-Star plant imported engines, transmissions, and so many other parts from Japan during a period when the yen had risen swiftly against the dollar, the costs for parts outstripped the revenue derived from the finished vehicles. In other words, Diamond-Star was losing money. Mitsubishi wasn't overly worried, because it planned eventually to buy more parts from American suppliers, thus cutting costs and making the assembly plant profitable. Besides, Mitsubishi's losses on finished vehicles, unlike Chrysler's, were balanced by the profit derived from parts sales. Chrysler's share of Diamond-Star's losses had amounted to $200 million since 1985. By selling its 50 percent stake back to Mitsubishi, Chrysler accomplished two objectives: it booked a $100 million profit on its initial investment and prevented further operating losses.

The sale of Chrysler's stake in the plant, coming on the heels of Iacocca's vitriolic attacks against Japan, was a step that widened the gulf between Chrysler and its Japanese partner. Toyoo Tate, the chairman of Mitsubishi Motors, was a patient man. He understood Iacocca's frustration with U.S. trade policy, which afforded Japanese automakers great freedom to operate in the U.S. market. But he also believed Chrysler's inability to sell cars was largely of its own making and could be rectified by improving its basic manufacturing. In its early days as a carmaker, Mitsubishi had learned so much from Chrysler. Now Chrysler was learning from Mitsubishi and had already used the knowledge to become a stronger competitor. To Mitsubishi executives, selling the Diamond-Star stake was a classic example of short-term thinking.

Tate had his own frustrations. He wished to strengthen the Mitsubishi brand in the consciousness of American carbuyers. Americans equated Japanese cars with Toyota, Nissan, and Honda. Mitsubishi, a giant conglomerate, sold merchandise as disparate as television sets and canned tuna. The Mitsubishi brand—represented symbolically by an array of three diamonds—didn't connote cars, and the name was difficult for Americans to pronounce, let alone remember. Chrysler and Mitsubishi had been partners since the early 1970s. According to their agreement, Mitsubishi had been allowed to sell small, fuel-efficient cars through Chrysler dealers under such brand names as Dodge Colt. But in the early 1980s, when Chrysler was floundering and needed financial help, Mitsubishi traded assistance to Iacocca and Chrysler in return for permission to sell vehicles in the United States under its own name.

Relations between the two companies, despite the strain, remained close. When Chrysler lacked six-cylinder engines for its popular minivans, it bought them from Mitsubishi. Still, Iacocca's attacks on Japan embarrassed and annoyed Tate. Every time Steve Miller or another Chrysler executive traveled to Japan, Tate produced newspaper clippings of Iacocca's attacks on Japan and the Japanese auto industry. He wanted to know why Iacocca openly blamed Japan for America's trade imbalance when Chrysler was relying heavily on Mitsubishi for cars, engines, and other imported parts. The Japanese goods ordered by Highland Park added to the very deficit that Iacocca constantly complained about. Did Chrysler have a special right to import from Japan that it denied to ordinary American consumers? Miller didn't really have an answer for the apparent contradiction except to plead for understanding that Iacocca's public bluster was in no way intended to insult Mitsubishi.

Iacocca's stance against Japan turned progressively more strident the more Chrysler's financial troubles grew. In March 1991 — less than six months after his attempt to sell Chrysler to Fiat failed — Iacocca wrote a private letter to George Bush, warning the president that unless car and truck imports from Japan were sharply limited for at least twelve to eighteen months, Chrysler might go out of business. The existence of Iacocca's letter to Bush came to light when the *Detroit Free Press* and the *Detroit News* filed a Freedom of Information request and obtained a Treasury Department briefing paper on automotive imports. "Even GM is at risk unless and until we decide as a nation that a domestic auto industry is a strategic industry and should not be allowed to fall victim to Japan's industrial strategy of targeting key industries," Iacocca wrote.

Though Iacocca was on better terms with George Bush than he had been with Ronald Reagan, Bush's administration had just as little sympathy for Chrysler and the rest of the auto industry. Ideologically, Bush favored free trade and wasn't inclined to protect industries that weren't willing or able to measure up to international competition. Moreover, administration officials privately regarded auto executives as gluttons during prosperity and whiners during economic downturns. The millions that Iacocca and other auto executives had showered on themselves during the 1980s and the questionable billion-dollar acquisitions they carried out, though not strictly germane to the question of trade policy with Japan, hurt Detroit's case with the administration's economic theorists. Companies and industries that sought Washington's help in trade matters should have displayed more frugality and stuck to their knitting. Instead of diversifying, Chrysler should have been finding ways to match Japanese competitors in its core business, cars. In any event, it hardly seemed like good policy to punish Japanese automakers; as detailed in another Treasury Department briefing paper, they had invested $6.4 billion in U.S. plants and employed tens of thousands of Americans.

Domestic politics conspired to alter the Bush administration's stance. In the fall of 1991 the administration embarked on what looked like a cakewalk to reelection in 1992. Bush's decision to crush Saddam Hussein's forces succeeded gloriously. Communism in Europe was breathing its last. At home, however, political storm clouds were gathering. The auto industry was reeling from a two-year slump, caused by weak demand for new vehicles and increasing penetration by Japanese brands. Democratic legislators were noisily proposing bills to limit Japanese market penetration, threatening to upset the administration's free-trade initiatives. At the planning meeting for a Far Eastern trade mission in January 1992, the proposal was made, and seconded, to invite the chairmen of the Big Three to accompany George Bush on one leg of his upcoming trip to Japan. With Big Three executives at his side, Bush could demonstrate toughness and leadership on domestic economic concerns. Bush's campaign advisers hoped he might extract some concession from Japan that would placate Detroit and Democrats in Congress while maintaining cordial economic relations with an ally.

The Far East trade mission was an epic disaster for Bush and the U.S. auto industry. General Motors provided an omen of what lay ahead when it announced in early December, a month before the trip, that it planned to close twenty-three factories and cut more than 70,000 jobs through 1995. Bush had been hoping to position himself as a champion of American jobs. As soon as the mission landed in Japan in early January, it was put on the defensive. The industry's position that American cars had been unfairly barred from Japan appeared silly in light of the numerous experts, Japanese and American, who were explaining the nature of the trade battle to journalists traveling with the president. The fact that only a handful of American models was available with right-hand drive, the Japanese standard, constituted immediate and damning evidence against Detroit in the minds of the media. Actually, the reasons why Japan wasn't buying American cars were much more complex than the instant analyses of Peter Jennings and others revealed. Japan, to be sure, was guilty of questionable trade practices. But the steering-wheel issue was simple enough for journalists and TV commentators to explain. Americans easily understood that no one, including the Japanese, was likely to buy cars with the steering wheel on the wrong side of the car. Bush and the chief executives of the auto companies appeared very foolish indeed for suggesting that unfair trade practices was the most important reason for trade deficits.

George Bush looked ridiculous because he had listened to advisers who never fully grasped the reasons for trade tensions between Detroit and Japan. Inviting Lee Iacocca, Red Poling, Bob Stempel, and Owen Bieber on the trip had been a massive blunder. The Detroit contingent has so often repeated the fairy tale that Japan's protected markets were the root of its troubles that it had actually begun to believe it. Of course Japan *did* erect barriers to the American car industry and other foreign automakers. And Japanese automakers sometimes violated trade rules. But the American car industry had not been much interested in Japan for years. Detroit's automakers said that they might have tried harder in Japan if they hadn't perceived the deck as being stacked against them, but they weren't able to produce much evidence that they had tried. Iacocca charged that the Jeep Cherokee, which Chrysler sold through Honda dealers in Japan, sold for $12,000 more in Japan than in the United States, supposed proof that the Japanese market was rigged. But Honda officials demonstrated that Cherokees had to be virtually rebuilt to meet Japanese quality standards. What Iacocca neglected to say was that the Cherokee's price in Japan had been set by Chrysler, not Honda.

Japan's "rigged market" was a convenient scapegoat for the trade deficit, much less painful than admitting that Japanese automakers had satisfied American customers better than Detroit had. Lee Iacocca needed Japan's "unfairness" as the premise for American trade actions against Japan. His ploy was rather transparent. As the *New York Times* opined in "The Excuse Maker," a stinging editorial about Iacocca's participation in Bush's trade mission, "The solution lies in making better cars in this country, not angrier excuses about Japan."

Bush and Detroit's three musketeers marshaled few compelling arguments to wring concessions from Prime Minister Miyazawa or the Japanese bureaucracy. To help the American president save face during his trip, however, the Japanese government had persuaded automakers in Japan ahead of time to collectively promise an additional $20 billion worth of automotive purchases from the United States. The promise wasn't a tough one to extract since additional purchases made financial sense anyway due to the weakness of the dollar and the improving quality of American-made merchandise. In any case, some of the additional purchases would come from U.S. factories owned by Honda, Toyota, and Japanese parts manufacturers.

On January 10, 1992, the day after the trade mission returned, Iacocca addressed an overflow crowd at the Economic Club of Detroit. He characterized the results of the mission as "too little, too late." In his most vitriolic attack ever, Iacocca fanned the flames of ethnic hatred by accusing Japan of attempting to colonize the United States with its economic policies and with the "insidious Japanese economic and political clout within the United States." He accused Toyota, which had been charged with selling minivans in the United States in violation of dumping laws, of "cheating" and of "breaking our laws," despite a judge's finding that exonerated the Japanese. He rehashed the fact that the Jeep Cherokee cost $12,000 more in Japan than in the United States. Unfairly high prices and red tape were the reasons Chrysler and other U.S. automakers weren't selling more cars there.

"They're beating our brains in," Iacocca wailed indignantly. That was the phrase that made the papers the next day.

"I used to believe myself that the Oriental 'long view' was a great virtue and something we could learn from the Japanese," Iacocca said. "I was wrong. It's not a virtue at all, it's a weapon, and we have to disarm them if we're going to get anywhere at all. We need to use our weapon, good old-fashioned American impatience. That means demanding a solution to the problem now. And retaliating now if we don't get it." He then lit into the press and editorialists for criticizing his stance against Japan and for ridiculing him and the other auto executives as "window dressing," "clowns," and "potted plants" during Bush's trade mission.

Finishing on an upbeat note for the hometown crowd, Iacocca delivered a plug for Big Three car and truck models to be introduced at the North American International Auto Show that would open the following week in the adjoining exhibition hall. The display included Chrysler's Jeep Grand Cherokee, which was being assembled just five miles away, on Jefferson Avenue, the Viper, and the L/H prototypes, set to be introduced later in the year.

Iacocca had been more worked up than usual at the humiliating way the auto industry was being portrayed. His speech epitomized the sadness and anger of a star whose triumphs were fading further and further into memory. It served, in a way, as a swan song, a final chance to cry out and flail against the forces that were transforming his world. Time had nearly run out. As he stood on the podium, Lee Iacocca knew that in a few months he must relinquish center stage to someone younger and stronger who could pursue a fresher, more relevant agenda than his. The power in the Motor City was moving to a generation with ideas and rules that were foreign to him.

THE TURNING POINT

THE FUTILE BUSH MISSION to Japan, which clumsily attempted to alter economic reality through trade bullying, foreshadowed failure of the president's reelection campaign and brought down the curtain on a generation of automotive leaders. Within months, GM's directors replaced Bob Stempel; and a year later, Red Poling would retire from Ford and Owen Bieber would negotiate his last labor contract as president of the United Auto Workers Union.

Lee Iacocca's tenure, too, was winding fitfully to a close. The automaker's board of directors, realizing it had blundered by allowing Lee to dominate the succession process and, therefore, stall for time, had forced him to announce his pending retirement a few months before the Bush trip. The last executive privilege that Iacocca retained was to nominate a successor—and the director's patience wasn't endless on that subject either. The candidacy of Roger Penske had fallen through. Jerry Greenwald was an unacceptable choice, because directors weren't willing to alienate Bob Lutz, who assured them he would quit if Jerry returned. Nor did directors think it fair that Jerry should be rewarded for having jumped ship. Steve Miller—never more than a long shot—was forced to retire by Iacocca because he dared to suggest that the boss had stayed on too long. Tom Gale and Jerry York lacked experience.

Lee Iacocca's career had, in the end, come down to a cliché that contained so much truth: it was lonely at the top. As Iacocca searched about, trying to figure out who was on his side and who was against him, he finally understood that the pinnacle of corporate power *was* a cold, frustrating and treacherous place—just as it unquestionably had been for his old nemesis, Henry Ford II. Perhaps now it was easier for Iacocca to understand the suspicions Henry had about him and about most of the people around him, and why Henry had had such difficulty in choosing someone whom he trusted to take his place. Iacocca was finally in a perfect position to see that in the thirteen years since getting fired, he had, in many ways, become Henry Ford.

It was more than just the cigars, the arrogance, and the jet-set lifestyle. In the old days things happened because Henry willed them to happen. That was the power and the position that Lee had aspired to forty-six years earlier as a freshly minted mechanical engineer from eastern Pennsylvania. He had achieved just that, rising to become a brilliant industrial leader who persuaded the government and the workers to give his company another chance and then leading it back from the edge of oblivion. That was the power he didn't want to give up. But he let himself get caught by the perils of celebrity, and he drifted too far from the nitty-gritty of automaking to comprehend how dramatically his dominion had been transformed during his reign. His inheritor would be a different breed of leader. Directors now wanted someone who didn't rule by decree but who used his power to guide, encourage, and stimulate.

"Who else is out there?" Iacocca asked Ben Bidwell. Bidwell, officially retired but still working at Chrysler as a consultant, was among the few people Iacocca trusted. Iacocca had always felt comfortable with Bidwell and could confide in him. Iacocca knew that if he didn't come up with a candidate soon the board was liable to act on its own. And the choice might be unpalatable to him.

"What about the GM guy, Eaton?" Bidwell asked.

"Where did you get that name?" Iacocca demanded to know. He was shocked because he had quietly been doing homework on Bob Eaton, and had forgotten that he and Bidwell had casually discussed Eaton a few months earlier when Roger Penske recommended him.

In December, Eaton's name came up again, and in a most unorthodox manner. Fred Hubacker, a midlevel financial executive at Chrysler, was worried, like many of his colleagues, that Chrysler was sliding toward a crisis as the company fiddled about trying to find a new leader. Hubacker, who knew through the grapevine that Lutz and Greenwald weren't viable candidates, decided there was no harm in commending his neighbor and friend for twenty years, Bob Eaton. The Hubackers and Eatons lived near each other in Birmingham, Michigan, and owned neighboring summer homes in northern Michigan, where the auto industry takes its ease. Their wives were teachers and friends. The men liked to ride dirt bikes together. Though several levels below Iacocca on the corporate ladder, Hubacker didn't see what harm could come from a letter to the chairman.

Incredibly, Iacocca responded to Hubacker's letter and asked him to arrange a meeting with Eaton.

Robert J. Eaton, a native of Buena Vista, Colorado, had studied mechanical engineering at the University of Kansas and climbed the ladder of GM's technical ranks. When GM was forced to defend the safety of the brakes on its front-wheel-drive X cars in the early 1980s during a government case, Eaton was the star witness. By 1986 he had been promoted to head of all technical staffs, in effect GM's chief engineer. From the technical center in Warren, Michigan, Eaton was promoted to president of European operations in 1988. His boss in Europe was Jack Smith, chairman of GM-Europe and a talented executive whom many believed to be a future GM chairman. When Jack was promoted to executive vice president in charge of all international operations, Eaton became chairman of GM-Europe.

Eaton's transfer to Europe provided him with exposure to non-Detroit thinking. Far removed from headquarters in Detroit, GM-Europe in many ways was free to operate as a free-standing automobile company with two major brands: Opel in Germany and Vauxhall in Great Britain. GM had posted massive deficits in Europe in the early 1980s, which Jack Smith had begun to reverse through massive cost-cutting and efficiency efforts, as well as through the development of more attractive, higher-quality vehicles. The profits generated by European operations in that turnaround sustained the company during a period of disastrous losses in North America.

Eaton was a low-key executive, and his style differed from that of the relentlessly autocratic Roger Smith. In the spirit of organizational reform that many companies were learning, he was willing to listen and to allow consensus to build before deciding an important issue. He was also quite willing to cut discussion short in the classic GM manner and take action when the situation required it.

The chance to run Chrysler that abruptly presented itself to Bob Eaton was attractive. At the age of fifty-two, and given the ages of his superiors at GM, he wasn't likely to have the chance to run his own company. Chrysler was less than half the size of GM, but to rule an automaker, a $50 billion enterprise, as chief executive, was a once-in-a-lifetime chance to make a deep and lasting mark on American business, to shape design, technology, labor relations, and manufacturing the way no second fiddle ever would.

Even from the considerable distance of GM-Europe's headquarters in Zurich, Eaton was all too aware of the politics and backbiting in which Chrysler was embroiled. One only had to read the newspapers and hear the industry gossip to realize that Lutz had run afoul of Iacocca and was being passed over. It was evident that Lutz—not Iacocca—was the force behind the new Chrysler vehicles. If Eaton was named CEO, Lutz might quit, or worse, might try to subvert Eaton in an attempt to fight his way into the chairman's job. Evidently, Iacocca, two years beyond normal retirement, was reluctant to step aside and was under pressure from directors. That Iacocca might try to control Chrysler from the boardroom, as nonexecutive chairman or some such, was not beyond the realm of possibility.

During the Christmas holiday of 1991 the Eatons flew home to Michigan. With Fred Hubacker playing the go-between, Iacocca and Eaton got together to chat. The meeting went well. After the holiday, three Chrysler directors—Mal Stamper, Bob Lanigan, and Jean de Grandpré—flew to Zurich to interview Eaton. Likewise, the directors were impressed with the man. They wanted to hear that Eaton favored the team approach to management. What Chrysler did not need was another egocentric leader or someone with grandiose tendencies.

Eaton flew secretly from Europe to San Francisco in late February 1992 to meet with Iacocca for a final interview. The conversation went exceedingly well, lubricated perhaps by the simple reality that Iacocca was out of options. Pressure from investors for a solution had built to a level that was just short of insurrection. John Neff, the manager of a prominent mutual fund that owned twenty million Chrysler shares, had called publicly for the immediate promotion of Lutz to chief executive. The directors and Iacocca *had* to produce a successor or face the consequences at the annual meeting.

Eaton left his meeting with Iacocca with a memorandum of understanding: subject to board approval, he was to join Chrysler immediately and take over as chairman and chief executive at the end of the year. He and everyone else, including Iacocca, were to try to convince Lutz to stay on as number two.

Iacocca, after retiring as chief executive, would be entitled to stay on the board for two years as chairman of the executive committee. He would be allowed to keep using Chrysler's jet. The arrangement was as close as Eaton could get to a promise from the board that Iacocca had no intention of keeping his hand on the tiller.

Before returning to Europe, Bob Eaton took a sidetrip to Cincinnati, home of Procter and Gamble and John Smale, P&G's former chairman. As the talks with Iacocca grew more serious, Eaton had felt obligated to inform Jack Smith and GM's chairman, Bob Stempel. Smale, an influential GM director, was not happy to hear about the possibility of losing Eaton to a competitor. Smale and other GM directors were dissatisfied with Stempel's leadership and planned to ride closer herd on him. If Stempel didn't cut it Jack Smith might be moving up, possibly as chief executive. Big things might be happening very soon, Smale told Eaton, and there will be opportunities for you. Smale wasn't able to show all his cards, but the implications were clear.

Smale's comments were tempting, but the fact remained that he and Jack Smith were about the same age — no matter what happened to Stempel, he was never going to run GM. When Eaton returned to Zurich in early March he tried futilely to hide the fact that he had decided to leave GM. Approval of his nomination by the Chrysler board, though not seriously in doubt, had not taken place and early disclosure might upset the apple cart. In answer to a reporter's questions, Eaton conceded that he had discussed the job with Iacocca. Rumors began to swirl, fueled by a front-page story in the *New York Times* reporting that Eaton was about to jump to Chrysler. But when an Associated Press reporter asked him during a ribbon-cutting at a new factory in Hungary if it was true that he was going to Chrysler, Eaton denied it. Shortly after the ceremony, Eaton was on an airplane bound for New York.

On Saturday, March 14, 1992, the Chrysler board, meeting at the Waldorf Towers in New York, approved Bob Eaton. The following Monday morning the newly elected vice chairman and successor to Iacocca's throne introduced himself to Chrysler's twenty-five top officers. Iacocca then introduced Eaton to reporters at the Highland Park headquarters, stating that the board had chosen him because he was "the most qualified to lead Chrysler." As Iacocca spoke, all eyes turned to Bob Lutz, who sat on the stage with Iacocca and Eaton.

For Lutz, the selection of Eaton was anticlimactic. He had known for months that Iacocca was blocking him, just as he himself had blocked Greenwald, Iacocca's choice. Faced with a stalemate, the directors had turned to a compromise candidate of Iacocca's choosing.

"Naturally I'm disappointed," Lutz told reporters. But he didn't plan to leave or to wage war against the new boss. "Being a team player means you don't sulk or quit when someone else is named captain of the team." Iacocca chimed in with gracious words for Lutz, describing as "crap" all the stories that the two executives had been at odds.

Iacocca's fib didn't bother Lutz. That's the way the game was played. If Lutz had any inclination to quit, it wouldn't be easy, given the $4 million price tag on his pending divorce settlement and the fact that his Chrysler options were nearly worthless due to the weak share price. First impressions often were misleading, yet Eaton immediately struck Lutz as the sort of CEO who would be fair and decent and acknowledge what Lutz and others had accomplished during the previous three years in order to change Chrysler and to create new and exciting vehicles.

The turning point came a few months later in July, at the Rancho Valencia Resort in California. There, in the foothills near San Diego, automotive reviewers and security analysts gathered to drive and evaluate the L/H, Chrysler's make-or-break car.

A few months earlier, Chrysler had introduced the Jeep Grand Cherokee, the first of its completely new vehicles since the advent of the platform team approach. The Grand Cherokee received respectful notice from reviewers and was selling briskly. But Chrysler didn't have the capacity to build more than 240,000 Grand Cherokees annually. From a financial perspective, the L/H was the far more crucial test, for Chrysler needed eventually to sell 400,000 or more L/H mid-size cars. Moreover the L/H, if successful, would provide the mechanical basis for a generation of future models.

The timing of the new vehicles was fortuitous. The cyclical U.S. auto market, which had been sluggish for nearly four years, was finally showing the first glimmerings of rejuvenation. No carmaker wanted to introduce new models to a weak market, as GM had done a year earlier. Chrysler couldn't afford a dud because its financial reserves were nearly depleted, and the banks were not enthusiastic about renewing their credit agreements. Memories of 1980, still very fresh and painful, loomed over the L/H's introduction. If the market weakened, if the L/H bombed, Chrysler might find itself seeking government aid once more.

The drivers behind the wheels of the L/Hs — which would be sold as Dodge Intrepid, Eagle Vision, and Chrysler Concorde — seemed quite pleased and surprised by what they experienced as they drove through the mountains toward Borrego Springs and the desert of southern California. The cars were quick and responsive; they reacted crisply to steering and acceleration commands; and their manners were refined over all sorts of pavement. The interiors were nicely appointed and the exteriors were actually rather hip: all in all, nothing like what Chrysler had been foisting on the public for the past decade.

At a barbecue that evening at Rancho Valencia, the consensus among the guests was rather one-sided: Chrysler, against all odds, had produced a very reasonable alternative to the Hondas, Tauruses, and other popular midsize sedans. Within hours, the security analysts were writing reports for investors about the stunning evidence of Chrysler's turnaround.

In July and August, positive comments about the L/H were appearing regularly in the press and in brokerage reports—and not a moment too soon. On August 18, 1992, a consortium of 152 banks agreed, after months of tense negotiations, to terms for a new $6.8 billion loan agreement with Chrysler Financial. At one point Chrysler Financial executives had appealed to Alan Greenspan, chairman of the Federal Reserve, to help convince the foreign banks to go along. Without the new loan agreement, Chrysler almost certainly would face a crisis of equal magnitude to its travails twelve years earlier. Once more the cavalry arrived in the nick of time.

The investiture of Bob Eaton and the acceptance by consumers of Chrysler's two newest vehicles were the signals, at long last, for the wave of farewell parties for Lee Iacocca. The biggest and gaudiest bash was the celebration held in Las Vegas, paid for by Chrysler dealers. Ben Bidwell served as master of ceremonies, an obviously tipsy Frank Sinatra crooned, and little was spared in the way of glitz or expense for the great man's send-off. Singers, dancers, and animal acts were engaged in a spectacle meant to rival Ben-Hur's chariot race.

The two Bobs, Lutz and Eaton, seemed to be managing Chrysler as harmoniously as the optimists had hoped. Eaton possessed the keen managerial sense and diplomatic skills to allow Lutz to enjoy the limelight for Chrysler's successful new vehicles. It didn't seem important to Eaton—perhaps because he was unusually secure about himself—to underscore that he was the new chief executive.

For Iacocca, Ford-sized dreams had run into 1990s reality. Lee was unhappy to relinquish active management, just as directors were uneasy about allowing him to stay, even in an advisory role. Eaton had been promised a clear mandate, which would be impossible if he was going to be second-guessed by an "elder statesman." Just as disturbing as the loss of power was the loss of perks. Lee hated the prospect of relinquishing the Gulfstream IV jet, the suites, the expense accounts, the limousine and podium that were waiting for him whenever he made a public appearance, and the ever-present retinue of assistants and go-fers who handled the details. Had it been easier for him to spend his own money he might have realized he had plenty to pay for his own jets and his own assistants. After all, a conservative accounting of the compensation he was paid at Chrysler suggested a personal fortune worth at least $100 million—the mere interest from which could keep him in style.

Iacocca's ace in the hole was Kirk Kerkorian, the billionaire financier who bought a 10 percent stake in Chrysler. While lounging on Kerkorian's yacht off Spain during the Olympics, Iacocca spoke of the shabby treatment he was receiving from the board in his negotiations for final separation. He complained the directors had promised to take care of him. Kerkorian wrote Eaton to say he was distressed about his friend's diminished role. The letter didn't have much effect except to stiffen the directors' resolve to finally wrap up negotiations. So, it was settled: while Lee remained on the board for two years he would be paid $500,000 per year under a consulting agreement. He was allowed to use the Gulfstream jet for company business and could lease it for personal use. The board also agreed to grant him some extra years of service so that he would qualify for a richer monthly pension—closer in league with the $1.1 million pension paid to Roger Smith when he left GM.

But, once more, Lee's ego became an issue. It was important for him to go out as top dog, and it was crucial that no one earn more than he or receive a bigger pension. So when Iacocca learned during his last months with the company that directors also added some months of service to Bob Lutz's pension—more than to his own—he insisted that his pension be adjusted upward once more. This time, the directors, several of whom were furious, wouldn't budge. Perhaps it was incredible—and, then again, perhaps not—that their refusal was followed soon after by a letter from an attorney representing Iacocca who threatened to sue unless his client's retirement benefits were sweetened.

Iacocca never made good on the threat. Less than a year into his retirement, bored and frustrated at serving on a board over which he had little influence, he resigned as a director and began spending more time on the West Coast, in Aspen, and in Palm Springs with his new wife, Darrien Earle. Because he no longer was regarded as an insider for purposes of federal securities law, he was able to sell a big chunk of the fortune in Chrysler stock he had amassed, and did so, netting $53.3 million.

Following the successful introductions of the Jeep Grand Cherokee and the L/H sedans, Chrysler introduced in 1993 its first new pickup truck in nearly twenty years and a new subcompact, the Neon; in 1994 came two new compact models, the Cirrus and Stratus. All the new cars received respectful reviews. The company's earnings soared and the price of the stock also rose sharply, rewarding many investors who had stood fast when prospects were bleak in 1990 and 1991.

Bob Eaton and Bob Lutz meshed well in the executive suite, dividing duties according to their strengths. Along with the rest of Chrysler they enjoyed the acclaim heaped on the automaker as the hottest of American car companies. Yet they couldn't help but be aware of some discomforting facts: Chrysler vehicles still were not at the top—or even close to the top—of the J.D. Power quality ratings surveys; the Japanese automakers were steadily expanding production in North America as a way to blunt the price-inflating effects of the strong yen; and, most important, the surging, outsized demand for cars and trucks that played so important a role in Chrysler's rebound was not going to last forever. How the people of Chrysler would perform in the stormy period that lay inevitably ahead would reveal a great deal about what they and the rest of Detroit's leaders truly had absorbed from the timeless lessons of their industry.

EPILOGUE

IN APRIL OF 1995 Chrysler, G.M., and Ford posted enormous first-quarter profits, continuing a two-year streak. In the automobile industry, strong financial results aren't necessarily a harbinger of good times; however, by early 1995 investors already were sensing a slowdown in the economy and began selling off their automotive stocks, driving share prices lower. According to the industry's historic pattern, when car and truck sales eased a bit, the stage would be set for the next recession.

Recessions have always spelled deficits for the Big Three and especially heavy losses for Chrysler. In 1995, however, Bob Eaton and Bob Lutz determined to break the pattern by controlling costs and saving cash to tide the company through lean times. Indeed, Chrysler's unit costs had become the lowest among domestic automakers. The low share price of Chrysler stock was irksome, but management hoped that investors would come back if the automaker could weather the next recession without government bailouts, stock issues, emergency loans or other heroic measures.

Kirk Kerkorian, whose 10 percent stake in Chrysler (worth about $2 billion) made him its largest investor, had other ideas. He watched his Chrysler shares peak at $63 per share and then decline to less than $40 a share over a period of approximately eighteen months. Kerkorian had bought most of his stake at about $10 a share, so the share value still represented a handsome profit for him. But Kerkorian hated watching his wealth decline at all. Because his stake was so large, dumping his shares would drive the price even lower.

In November, 1994, Kerkorian informed Eaton he wanted Chrysler to take strong action to increase the price of Chrysler's shares. In his view, Chrysler ought to use its cash to buy back stock — precisely the tack Iacocca had tried in the 1980s, with dismal results — or to raise the dividend.

To mollify Kerkorian, Eaton agreed in November to a modest share buyback program. Chrysler also raised its dividend slightly. The moves didn't impress Kerkorian or other investors: Chrysler's stock price continued to slide.

On the morning of April 12, 1995, Kerkorian shocked the financial world and the automobile industry by announcing an unsolicited bid to buy all of Chrysler at $55 a share, a transaction that would be worth close to $23 billion. Almost as breathtaking as the bid itself was the manner in which it was carried out. Kerkorian flouted Wall Street convention by tendering his offer without securing bank financing or the commitment of equity partners in advance. He did not even hire an investment banker to represent him. Under his plan, up to $5.5 billion of Chrysler's own cash would be used to secure borrowings, an idea pioneered by the leveraged buyout specialists of the 1980s.

Through his spokesman, Kerkorian explained that he was counting on the cooperation of Chrysler managment, a presumption that Eaton immediately branded as ludicrous. Chrysler saw Kerkorian's bid as a direct threat to the company's stability not to mention its 100,000 employees and the communities and institutions that depended on it.

Alex Yemenidjian, Kerkorian's top aide, insisted that he and Chrysler executives had been discussing a buyout for months prior to the bid. Gary Wilson, co-chairman of Northwest Airlines and an acquaintance of Kerkorian's, acknowledged that he, too, had discussed a highly leveraged buyout of Chrysler with Tom Denomme, vice chairman, and Gary Valade, chief financial officer. The would-be buyers had argued that Chrysler didn't need the $7.3 billion in cash that Eaton has amassed in anticipation of the recession. A big chunk of that money ought to be returned to shareholders through a share buyback or by a dividend increase. Denomme and Valade didn't deny that discussions had taken place; in their view, however, the discussions had been sketchy and noncommittal.

The most stunning feature of Kerkorian's bid was not its suddenness or its unconventionality but the role of Lee Iacocca. In 1990, when Kerkorian first bought Chrysler shares, the two men hardly knew one another. During the next four years they became allies; both lived in Los Angeles and spent time together in Las Vegas. Iacocca, who was a director of Kerkorian's MGM Grand, Inc. began investing in gaming ventures. According to Darrien Iacocca, he seethed over his forced exit from Chrysler and the fact that Eaton and Lutz seemed to be managing the company splendidly without him.

To assist Kerkorian's buyout, Iacocca agreed to invest his remaining $50 million worth of Chrysler stock in the buyout, small beer compared to the billions in debt and new equaity that would be required. His $50 million was not the essential ingredient. Because Kerkorian knew next to nothing about automaking, including Iacocca in his bid was meant to reassure lenders and potential partners that the would-be buyers had the experience to leverage assets and seize control of the nation's third largest automaker.

But Kerkorian may have badly overestimated Iacocca's good will with Chrysler shareholders. Eaton, Lutz, senior executives, Chrysler directors, and middle managers throughout the company dismissed the idea that Iacocca had anything positive to contribute to Chrysler. Newspaper editorialists, pension fund managers, retirees, and union leaders openly blistered Iacocca for presuming that his leadership was needed or wanted.

A few newspapers and magazines, struck by the apparent hypocrisy of Iacocca's participation, ran passages verbatim from his writings that had criticized the buyout specialists of the 1980s as vultures who stripped companies while promising to add value.

As Iacocca clumsily tried to explain himself, he only made matters worse. In one newspaper interview he criticized the quality of Chrysler's vehicles and the board of directors that had enriched him so handsomely. Kerkorian must have realized that Iacocca's participation had become a disastrous liability. Within a week of the announcement, Iacocca's public statements ceased.

From a technical standpoint, Kerkorian's attempt to put Chrysler in play was, in a way, as brilliant as it was hasty in its conception. He had risked nothing. Floating a highly conditional offer to buy Chrysler's shares didn't cost him a penny in loan commitment fees or investment banking charges. If his offer panicked Bob Eaton into a big share buyback, or attracted a rich foreign automaker to buy Chrysler, or prompted a shareholder uprising—then Kerkorian might have been able to sell his shares at a higher price. Otherwise, the worst that could happen would be that his shares would return to the April 11 price.

For Iacocca, the risk was somewhat different. Beyond his fortune, estimated at somewhere between $100 million and $200 million, his main asset was his reputation. Winning control of Chrysler could have put him back in the driver's seat with an influential position on the board, an opportunity to control events and to earn generous consulting fees, and perhaps a chance to resume his performance in Chrysler commercials as an experienced elder statesman. By joining a dubious venture to plunge Chrysler deeply in debt for the sake of a higher stock price in the short term, Iacocca relinquished the loyalty and the esteem of many who had regarded him, correctly or not, as a giant figure in American automaking.

There had always been at least two Lee Iacoccas: one, the smart, powerful auto executive famous throughout the world as the man who created the Mustang, refurbished the Statue of Liberty, and saved Chrysler; the other, a man viewed by some as a self-centered, money-driven executive, quick to make excuses for setbacks, a man, according to one colleague, "who always knew how to get to the head of any parade." By virtue of Kerkorian's foray, the latter Iacocca was presented to the world.

ACKNOWLEDGEMENTS

A REPORTER who covers a beat for a newspaper or magazine learns much more than he or she can cram into routine stories and features. This book is a compilation of much of what I learned—but have not had a chance to write about in depth—on my beat. The book covers Detroit, automaking, Lee Iacocca and Chrysler, international trade, manufacturing, and the executive culture of U.S. corporations during the ten years I have worked as a correspondent for the *Wall Street Journal,* the *New York Times* and, lately, as a columnist for the *Detroit Free Press.*

Because *Behind the Wheel at Chrysler* confronts the shortcomings as well as the strengths of the U.S. auto industry and redefines the myth of Lee Iacocca, I expected and received minimal official cooperation from Chrysler and from Iacocca. For the record, Iacocca met with me once for half an hour and then declined further interviews. However, several members of Chrysler's management and officials from its public relations department offered unofficial guidance, granted interviews, and answered specific questions. In addition, several Chrysler executives granted interviews privately, that is, without guidance or arrangement by the company. They are acknowledged in the notes following the text. I thank them from the bottom of my heart, since this book would have been impossible without their generous assistance.

Numerous people gave unstintingly of their time to help me understand my subject. Among those who deserve special thanks are David E. Davis, Leon Mandel, Jim Womack, Allan Gilmour, Red Poling, Don Peterson, John McElroy, J.T. Battenberg, Ben Bidwell, Jerry Greenwald, Hal Sperlich, Darrien Iacocca, Peggy Iacocca, Dave Power, Carroll Shelby, Doug Fraser, Frank Joyce, Steve Harris, Baron Bates, Paul Lienert, Steve Scharf, John Wolkonowicz, Joe Campana, Gloria Lara, Ron DeLuca, Laurel Cutler, Maggie and Steve Miller, Fred Zuckerman, Judith Muhlberg, Joe Cappy, Walter Hayes, Shinichi Tanaka, Jeff Leestma, Shigeaki Kato, Jim Olsen, Bob McCurry, Roger Lambert, Dick Recchia, Rick Lepley, Clark Vitulli, Tom McDonald, Ted Nagase, Bob Hall, Dennis Gormley, Heinz Prechter, Martin Swig, Keith Crain, Jerry Paul, Susan Jacobs, Maryann Keller, Dave Healy, John Casesa, Ron Glantz, Wendy Needham, Harvey Heinbach, Jack Kirnan, John Koenig, Tim Andree, and Steve Schlossstein. The list of every authority who deserves acknowledgement and thanks obviously is longer. To all of you: my sincere gratitude.

Robert, Alan, and Victor Potamkin, owners of the Potamkin automobile dealerships, were particularly helpful with explanations of the automobile business from the retailer's perspective. Robert predicted long ago that Chrysler would bounce back, and, as in so many things, he was right. Thanks to all of you.

Tomiyuki Okusa of the Mitsubishi Motor Corporation helped me to arrange my travel and research in Japan. It is no exaggeration to say that what turned out to be a wonderfully rich and informative three weeks in Japan might have been a disaster without his help and advice. While I was in Japan I was received warmly by Ted Nagase of Mazda, Steve Berkov of Toyota, and in the homes of Jake Schlesinger and Louisa Rubinfein and Andy and Kasumi Pollack, fellow reporters working for, respectively, the Tokyo bureaus of the *Wall Street Journal* and the *New York Times*. Thanks to all of you.

Maggy Ralbovsky, Noam Neusner, and Jill Davidson Sklar assisted me in compiling background material. Maurena Muldoon helped me manage the constant flow of letters, packages, and telephone calls. Laura Berman, a skilled journalist and friend, read the manuscript and suggested changes.

Chip Visci, Tom Walsh, Bob Magruder and Heath Meriwether of the *Detroit Free Press* helped and supported this book during my first months of employment at the newspaper when I was trying to figure out how to write a column as well as finish this book. Many thanks to you.

My literary agent, Jane Gelfman, has been a constant source of encouragement and ideas; she is a terrific ally, confidant, and friend. Cork Smith and Yoji Yamaguchi, my editors at Harcourt Brace, helped me at the beginning of this project to focus my ideas and, toward the end, to express them as well as I could.

Whenever quotes are used I was able to confirm the substance of a conversation from at least two sources. When I didn't feel confident to quote precisely, I paraphrased. Naturally, the responsibility for any errors is mine.

NOTES

1. SUNSET

Late August 1991 marked the beginning of a six-month period during which directors of the Chrysler Corporation asserted their authority over the process of finding a successor to Lee Iacocca. Iacocca's reaction to the directors' actions were related by several friends and business associates as well as by directors themselves.

Two directors, Jean de Grandpre and Paul Sticht, granted interviews in the late summer of 1994 that detailed some of the board's reasons for asserting its authority and the events that followed. Mr. Sticht agreed to be interviewed at his private office in Meredith, N.H., and Mr. de Grandpre at a hotel in Montreal.

Chrysler officials declined to comment on the search, and Mr. Iacocca repeatedly declined requests for interviews. Bob Lutz, Steve Miller, and Jerry Greenwald, all of whom were candidates to succeed Iacocca, and Ben Bidwell agreed to be interviewed.

2. EIGHTY YEARS OF SUPREMACY

The history of the Chrysler Corporation and the first American automakers through the early 1960s is well documented in a variety of standard texts. I relied on Reich and Donahue's *New Deals* and Moritz and Seaman's *Going for Broke*—both about the bailout—which have extensive material on Chrysler's history.

From the 1960s on, Chrysler's history was covered extensively by *The New York Times*, *Time*, *The Detroit Free Press*, and other periodicals, from which I drew a great deal of material.

3. THE REAL LEE

Iacocca's autobiography, Lacey's *Ford: The Men and the Machine* and Halberstam's *The Reckoning*, about Ford and Nissan, provided a great deal of material about Iacocca's life and early career. Many of the anecdotes were verified by Ford public-relations officials and middle managers who worked for Iacocca and with him. Bill Fugazy, a friend of Iacocca who sold travel services to Ford, and Don DeLuca, an advertising executive and longtime associate of Iacocca, granted interviews in 1992 and 1993. Hayes's *Henry*, about Henry Ford II, contained helpful details and analyses.

4. BAILING OUT

New Deals and *Going for Broke* provided invaluable insight into the negotiations among Chrysler, the U.S. government, the United Auto Workers union, suppliers, bankers, and other parties interested in the automaker's near failure in 1979 and 1980.

Steve Miller and Jerry Greenwald, both of whom were instrumental in negotiating the Government Loan Guarantee Act, explained the events and deepened my understanding of the personalities involved. Douglas Fraser, president of the UAW during the bailout, explained the union's role.

Articles published in *The New York Times, The Wall Street Journal, The Detroit Free Press*, and other periodicals covered this era in great factual detail.

5. THE SEEKERS

Douglas Fraser, former president of the UAW, granted interviews to discuss the role of the union in helping Chrysler to recover. James Womack, a manufacturing consultant, and David Power, founder of J. D. Power & Associates, have given unstintingly of their time in the past several years to provide background on the auto industry's manufacturing efficiency and on vehicle quality.

Jay Chai, executive vice president in the United States for Itochu (previously C. Itoh), explained the history of the first and second automotive industry studies carried out at MIT.

6. PERILOUS PROSPERITY

Several Big Three financial executives — including Steve Miller, at Chrysler and Allan Gilmour, at Ford — were kind enough to explain the dynamics of cash flow, investment, and profitability during the cycles of vehicle buying.

Harold Sperlich, Chrysler's former president, described the automaker's product strategy during the early days of the government-loan guarantee, and his own role, later, as president.

The impact of Iacocca's autobiography on the book market and the auto industry was documented extensively by articles in *The New York Times*.

The author's notes and recollections from Iacocca's press conferences and interviews, starting in 1984, provide the basis for much of this chapter.

A member of the Statue of Liberty fundraising committee described the work of the committee, Iacocca's role, and the conflict with former Treasury Secretary Donald Regan.

7. THE GILDED AGE

Several former Chrysler executives — Miller, Bidwell, Fred Zuckerman, and others — granted interviews to discuss Chrysler's diversification strategy of the mid-1980s. Chrysler's acquisitions — particularly of American Motors — were reported on at length by the financial press, especially *The Wall Street Journal* and *The New York Times*.

I wrote several articles for *The New York Times* concerning the overproduction of vehicles, how that hurt the automakers, and the defection of younger buyers to foreign brands. Heavy sales of cars to daily rental companies and the phenomenon of "almost new" cars' eating into new-car sales first appeared in the popular press in *The New York Times*.

8. THE STORY OF BOB LUTZ

Bob Lutz granted several extensive interviews between 1990 and 1994 — some for newspaper articles in *The New York Times* and some for this book — on his background and career. Walter Hayes, Ford's former public-relations chief, and numerous current and former Ford executives discussed Lutz's history at Ford and his decision to join Chrysler.

Halberstam was the first to write about Lutz's role in the automobile business. According to Lutz, Halberstam gave too much credence to those who told him that Henry Ford II had been disturbed by the amount of publicity focused on Lutz. In the fall of 1994 interviews with Henry Ford II by David Lewis, of the University of Michigan, in which Ford praised Lutz's abilities, were made public.

9. JERRY GREENWALD'S STORY

After leaving Chrysler in 1990, Jerry Greenwald granted me several interviews on his background, his career at Ford and Chrysler, and his decision to leave Chrysler.

10. GODZILLA ECONOMICS

Toyota: A History of the First Fifty Years, published by the automaker, traced the history of the Toyota family from their first efforts to export cars to the United States. James Olson, Toyota's senior vice president in charge of public affairs, was helpful in providing the company's perspective on trade politics.

Jagdish Bhagwati, of Columbia University, and Gary Saxonhouse, of the University of Michigan, were very helpful with comments on the theories underlying government trade policy and the interests of the automakers.

New Deals and *Going for Broke* were key to understanding the government's and Chrysler's differing perspectives on what types of aid would be extended.

11. THE NAGOYA RULES

It would have been impossible to understand (or explain) the essential differences between the Toyota Production System and the system used by the Big Three without MIT's seminal work and the book that followed, *The Machine That Changed the World*. Additionally, co-author James Womack was extremely helpful in amplifying many of the points made in that book.

Alfred Sloan's memoirs were vital, not only to understanding how and why General Motors organized itself as it did but also to understanding the rise of professional managers.

The "five whys" method, developed by Toyota to find the root cause of problems, was explained to me during a visit to Toyota's offices in Torrance, California, and on a subsequent visit to Toyota's assembly plant in Georgetown, Kentucky. Discussions with Toyota executives illuminated the company's distinctive method of dealing with suppliers.

Clark and Fujimoto's *Product Development Performance* contributed greater understanding of the importance of a "large project leader" for vehicle-development teams.

12. ASIA IN AMERICA

The events and outcry surrounding the beating death of Vincent Chin in 1982 were reported in great detail by wire services, national newspapers, and *The Detroit Free Press* and *Detroit News*. By using the Nexis data base I was able to reconstruct a narrative of the incident, the criminal proceedings, and the ensuing protests by Asian-American groups.

The Asian-American Journalists Association of Los Angeles provided analysis of the propagation of negative stereotypes of Japanese business in its annual critiques of news media coverage of Asian Pacific Americans.

The U.S. Department of Commerce and Nippon Steel provided helpful data pertaining to merchandise and currency trade balances between the United States and Japan and between those countries and the rest of the world.

The Japanese automakers provided production data that showed the steady rise of transplant production in the United States since Honda's first plant in 1982.

The stances of the Big Three automakers regarding their lobbying positions in Washington were reflected in interviews throughout the 1980s and early 1990s with Marina Whitman, vice president and group executive for GM, Peter Pestillo, executive vice president of Ford, and Rob Liberatore, a vice president of Chrysler.

The diplomatic history of relations between the United States and Japan was drawn from the skillful and fluent *Let the Sea Make a Noise* by Walter A. McDougall. Toyota's official company history and Michael Cusamano's study of management and technology at Nissan and Toyota, *The Japanese Automobile Industry*, provided the background for the history of the Japanese auto industry.

Observations about how the history of Japan helped define the size and configuration of Japanese cars were drawn heavily from general reference material, as well as interviews with government officials and executives of auto companies during a trip to Japan in October 1993.

The passages about Japanese sensibility in automaking were based on the speeches of Kenichi Yamamoto, former president of Mazda Motor Corporation, and on a series of English-language books on Japanese aesthetics published by Mazda.

13. THE HONDA TEAM

The passages about Detroit losing young buyers grew out of a story I wrote on the subject in *The New York Times* in 1991.

Information from interviews with Harold Sperlich, former Chrysler president, on how the Honda team was formed was confirmed by several current and former Chrysler officials and several members of the team. Steve Miller, who hired Reiko McKendry, described her role.

The results of Chrysler's Honda team and conclusions drawn from them were compiled in an internal study provided to me by an official of Honda.

14. DREAMS MEET REALITY

Extensive accounts of the marital problems of Iacocca and Peggy Johnson ran in both Detroit newspapers; additionally, she explained much of what happened to her in an extraordinarily frank interview on Geraldo Rivera's television show.

Bob Lutz and Francois Castaing explained in separate interviews how they became allied and why they sympathized with one another. David E.

Davis, in addition to his published comments in *Automobile,* explained what he knew about Chrysler's product-development process, gleaned from his own diligent reporting on the company.

I attended the introduction of the Imperial in November 1989 and interviewed Joe Campana, who was head of the Chrysler-Plymouth division at the time and retired shortly after.

15. THE MAKING OF THE L/H

Chrysler afforded press and public a chance to scrutinize thoroughly the development of the L/H midsize car. This resulted in numerous stories in the mainstream and trade press. Chrysler's openness stemmed from the fact that in 1990 and 1991 there was considerable doubt about whether the automaker would be able to stay in business. Its public-relations department suggested showing writers and analysts exactly what was happening — normally, closely guarded secrets — in order to convince investors and car buyers that Chrysler was a viable enterprise. The material in this chapter was drawn chiefly from personal observation, as well as stories in *Automotive News* and other publications.

Dennis Pawley, executive vice president, and other manufacturing executives granted interviews about the L/H, the Jeep Grand Cherokee, and management changes during that period.

16. THE CHANGING OF THE GUARD

Mike Hammes and Joe Cappy related their recollections of the negotiations with Fiat. In late 1990, at about the time these were taking place, Iacocca gave a wide-ranging interview to *The Wall Street Journal* in which he conceded that he had made numerous management errors, including the attempt to diversify.

Ben Bidwell explained his role in the attempt to recruit Roger Penske to be Iacocca's successor.

Iacocca's anti-Japan rhetoric was a prominent feature of the speeches he gave and newspaper articles printed under his name during the commemoration of the fiftieth anniversary of the attack on Pearl Harbor.

17. ENDGAME

Doug Anderson, the management consultant who coordinated the Kohler meetings, described the executive conference at West Palm Beach, Florida, at which Iacocca agreed to scale back the technology center.

Glenn Gardner, executive in charge of the L/H product team, granted several interviews during the car's development and permitted me to attend a planning meeting.

Bob Lutz, Francois Castaing, and several other Chrysler executives and managers granted interviews describing the tension between Iacocca and Lutz during the development of the L/H.

Jean de Grandpre gave me an interview on events pertaining to the board's decision to take over the succession process in 1991.

18. THE TURNING POINT

Fred Hubacker, a friend of Bob Eaton's and a former Chrysler executive, explained his role in the recruitment of Eaton as chief executive.

I attended the preview of the L/H cars at the Rancho Valencia resort in July 1992.

BIBLIOGRAPHY

Altshuler, Alan, Martin Anderson, Daniel T. Jones, Daniel Roos, and James Womack. *The Future of the Automobile.* Cambridge, MA: MIT Press, 1984.

Asobi: The Sensibilities at Play. Edited by Mitsukuni Yoshida and Sesoko Tsune. Hiroshima, Japan: Mazda Motor Corporation, 1987.

Bhagwati, Jagdish. *Protectionism.* Cambridge, MA: MIT Press, 1988.

Clark, Kim B., and Takahiro Fujimoto. *Product Development Performance.* Boston: Harvard Business School Press, 1991.

Compact Culture, The. Edited by Mitsukuni Yoshida, Sesoko Tsune, and Ikko Tanaka. Hiroshima, Japan: Mazda Motor Corporation, 1982.

Crystal, Graef S. *In Search of Excess.* New York: W. W. Norton, 1991.

Culture of Anima, The. Edited by Mitsukuni Yoshida, Sesoko Tsune, and Ikko Tanaka. Hiroshima, Japan: Mazda Motor Corporation, 1985.

Cusamano, Michael A. *The Japanese Automobile Industry: Technology and Management at Nissan and Toyota.* Cambridge, MA: Harvard University Press, 1985.

Dauch, Richard E. *Passion for Manufacturing.* Dearborn, MI: Society of Manufacturing Engineers, 1993.

Fucini, Joseph J., and Suzy Fucini. *Working for the Japanese.* New York: The Free Press, 1990.

Halberstam, David. *The Reckoning.* New York: Avon Books, 1987.

Hamper, Ben. *Rivethead.* New York: Warner Books, 1991.

Hayes, Walter. *Henry.* New York: Grove Weidenfeld, 1990.

Hybrid Culture, The. Edited by Mitsukuni Yoshida, Sesoko Tsune, and Ikko Tanaka. Hiroshima, Japan: Mazda Motor Corporation, 1984.

Iacocca, Lee, with Sonny Kleinfeld. *Talking Straight.* New York: Bantam Books, 1988.

Iacocca, Lee, with William Novak. *Iacocca: An Autobiography.* New York: Bantam Books, 1984.

Ingrassia, Paul, and Joseph White. *Comeback.* New York: Simon & Schuster, 1994.

Lacey, Robert. *Ford: The Men and the Machine.* New York: Ballantine Books, 1987.

Martineau, Lisa. *Caught in a Mirror.* London: Pan MacMillan Publishers, 1993.

McDougall, Walter A. *Let the Sea Make a Noise.* New York: Basic Books, 1993.

Morita, Akio, and Shintaro Ishihara. *The Japan That Can Say No.* Unauthorized English translation that was circulated among U.S. lawmakers in 1989.

Moritz, Michael, and Barrett Seaman. *Going for Broke.* New York: Doubleday, 1984.

Naorai: Communion of the Table. Edited by Mitsukuni Yoshida and Sesoko Tsune. Hiroshima, Japan: Mazda Motor Corporation, 1989.

Petersen, Donald E. *A Better Idea: Redefining the Way Americans Work.* Boston: Houghton Mifflin, 1991.

Reich, Robert B., and John D. Donahue. *New Deals.* New York: Times Books, 1985.

Sakiya, Tetsuo. *Honda Motor: The Men, the Management, the Machines.* Tokyo: Kodansha International, 1982.

Samuelson, Paul. *Economics.* Tenth edition. New York: McGraw-Hill, 1976.

Sanders, Sol. *Honda: The Man and His Machines.* Tokyo: Charles E. Tuttle, 1975.

Schlossstein, Steven. *Trade War.* New York: Congdon & Weed, 1984.

Sloan, Alfred P., Jr. *My Years with General Motors.* New York: Doubleday, 1963.

Stuart, Reginald. *Bailout.* South Bend, IN: 1980.

Toyota: A History of the First Fifty Years. Toyota City, Japan: Toyota Motor Company, 1988.

Tsu Ku Ru: Aesthetics at Work. Edited by Mitsukuni Yoshida, Ikko Tanaka, and Sesoko Tsune. Hiroshima, Japan: Mazda Motor Corporation.

Womack, James, Daniel T. Jones, and Daniel Roos. *The Machine That Changed the World.* New York: Rawson Associates, 1990.

Wyden, Peter H. *The Unknown Iacocca.* New York: William Morrow, 1987.

INDEX

345